GCSE Success

WORKBOOK

Biology

Carla Newman
Joanne Barton

Contents

Revised

A Balanced Diet

Multiple-choice questions

Choose just one answer: A, B, C or D.

1. What are carbohydrates made of? **(1 mark)**
   - **A** glycerol
   - **B** glucose
   - **C** amino acids
   - **D** fatty acids

2. Which substance is needed for growth and repair? **(1 mark)**
   - **A** carbohydrate
   - **B** vitamins
   - **C** fats
   - **D** protein

3. Kwashiorkor occurs due to a lack of which substance? **(1 mark)**
   - **A** protein
   - **B** carbohydrate
   - **C** vitamin C
   - **D** fibre

4. How many main food groups are there? **(1 mark)**
   - **A** 5
   - **B** 6
   - **C** 7
   - **D** 8

5. Which of these substances is stored as adipose tissue? **(1 mark)**
   - **A** carbohydrates
   - **B** minerals
   - **C** fats
   - **D** protein

Score / 5

Short-answer questions

1. **True or false?** (4 marks)

	True	False
a) Proteins can be stored in the body.	☐	☐
b) First class proteins contain all the essential amino acids.	☐	☐
c) Plant proteins are examples of first class proteins.	☐	☐
d) Proteins are made of fatty acids.	☐	☐

2. List four factors that may affect a person's diet. (4 marks)

   i) ....

   ii) ....

   iii) ....

   iv) ....

Score / 8

GCSE-style questions

Answer all parts of all questions. Continue on a separate sheet of paper if necessary.

1 **a)** Explain what is meant by the term 'balanced diet'. **(1 mark)**

....

....

b) In terms of specific nutrients, what might a vegetarian lack in their diet? **(1 mark)**

....

2 Complete the following sentences. **(8 marks)**

Fats are composed of .... and ....

They are used by the body as an energy ....

Iron is an example of a .... It is needed to produce ....

Vitamin C prevents the disease called ....

Fibre is composed of ...., which helps prevent ....

3 Estimated average requirement of protein can be calculated using the following formula

EAR (g) = 0.6 × body mass (kg)

a) Calculate the EAR for a woman weighing 60 kg. **(1 mark)**

b) The woman becomes pregnant. Explain how this will affect her EAR for protein and how she will need to change her diet. **(3 marks)**

....

....

....

Score / 14

How well did you do?

0–6	Try again	7–12	Getting there	13–19	Good work	20–27	Excellent!

For more information on this topic, see pages 4–5 of your Success Revision Guide.

Homeostasis 1

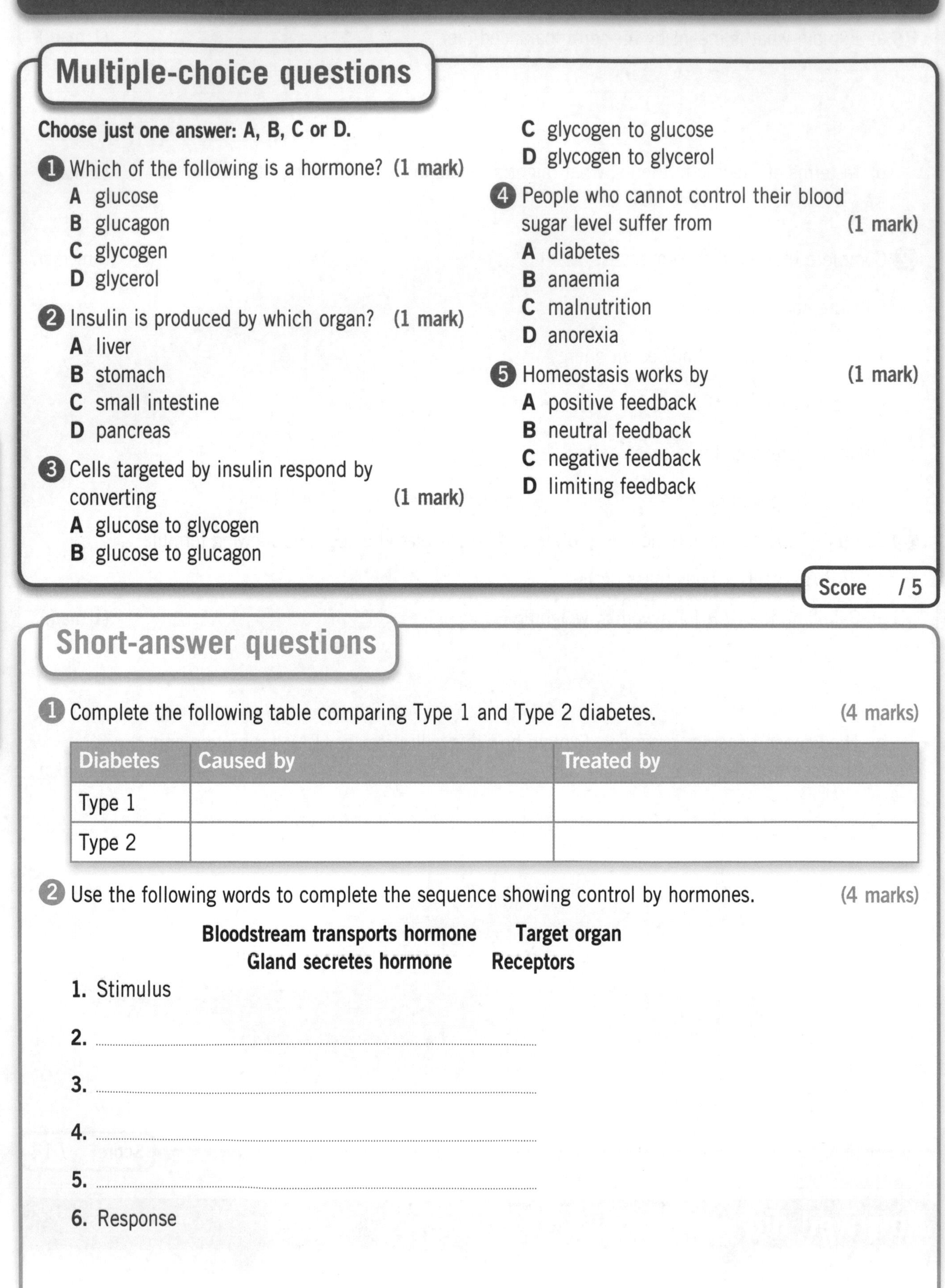

Multiple-choice questions

Choose just one answer: A, B, C or D.

1. Which of the following is a hormone? **(1 mark)**
   A glucose
   B glucagon
   C glycogen
   D glycerol

2. Insulin is produced by which organ? **(1 mark)**
   A liver
   B stomach
   C small intestine
   D pancreas

3. Cells targeted by insulin respond by converting **(1 mark)**
   A glucose to glycogen
   B glucose to glucagon
   C glycogen to glucose
   D glycogen to glycerol

4. People who cannot control their blood sugar level suffer from **(1 mark)**
   A diabetes
   B anaemia
   C malnutrition
   D anorexia

5. Homeostasis works by **(1 mark)**
   A positive feedback
   B neutral feedback
   C negative feedback
   D limiting feedback

Score /5

Short-answer questions

1. Complete the following table comparing Type 1 and Type 2 diabetes. **(4 marks)**

Diabetes	Caused by	Treated by
Type 1		
Type 2		

2. Use the following words to complete the sequence showing control by hormones. **(4 marks)**

Bloodstream transports hormone   **Target organ**
Gland secretes hormone   **Receptors**

1. Stimulus

2. ....

3. ....

4. ....

5. ....

6. Response

Score /8

GCSE-style questions

Answer all parts of all questions. Continue on a separate sheet of paper if necessary.

1 Define homeostasis, giving appropriate examples. **(3 marks)**

....

....

....

2 The following graph shows the blood sugar level of a diabetes sufferer.

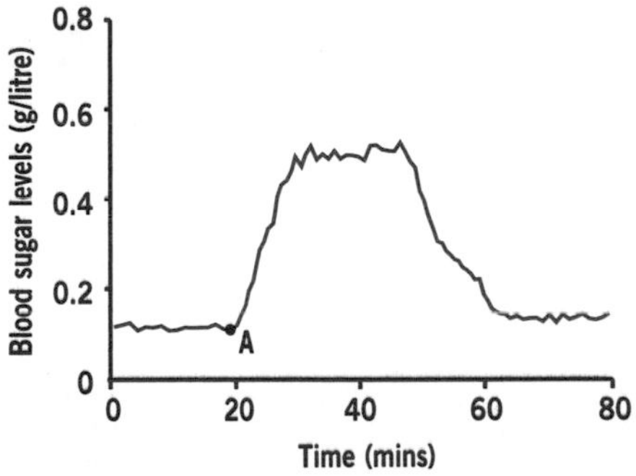

Blood sugar levels (g/litre)
0.8
0.6
0.4
0.2
0
A
0 20 40 60 80
Time (mins)

a) What is the normal blood sugar level of this person? **(1 mark)**

....

b) What caused the blood sugar level to increase at point A? **(1 mark)**

....

c) The person controls their diabetes by injecting insulin. At what time did they give themselves the injection? **(1 mark)**

....

d) Which type of diabetes does the person suffer from if they control it by injecting insulin? **(1 mark)**

....

e) A healthy person, who does not have diabetes, exercises for an hour. Their blood sugar level decreases below the normal level. Explain how the control mechanism returns the blood sugar level to normal. (You may use flow diagrams.) **(4 marks)**

....

....

....

....

3 Temperature control in a house works by negative feedback. Explain what this means and how it works. **(3 marks)**

....

....

....

Score / 14

How well did you do?

0–6	Try again	7–12	Getting there	13–19	Good work	20–27	Excellent!

For more information on this topic, see pages 6–7 of your Success Revision Guide.

Homeostasis 2

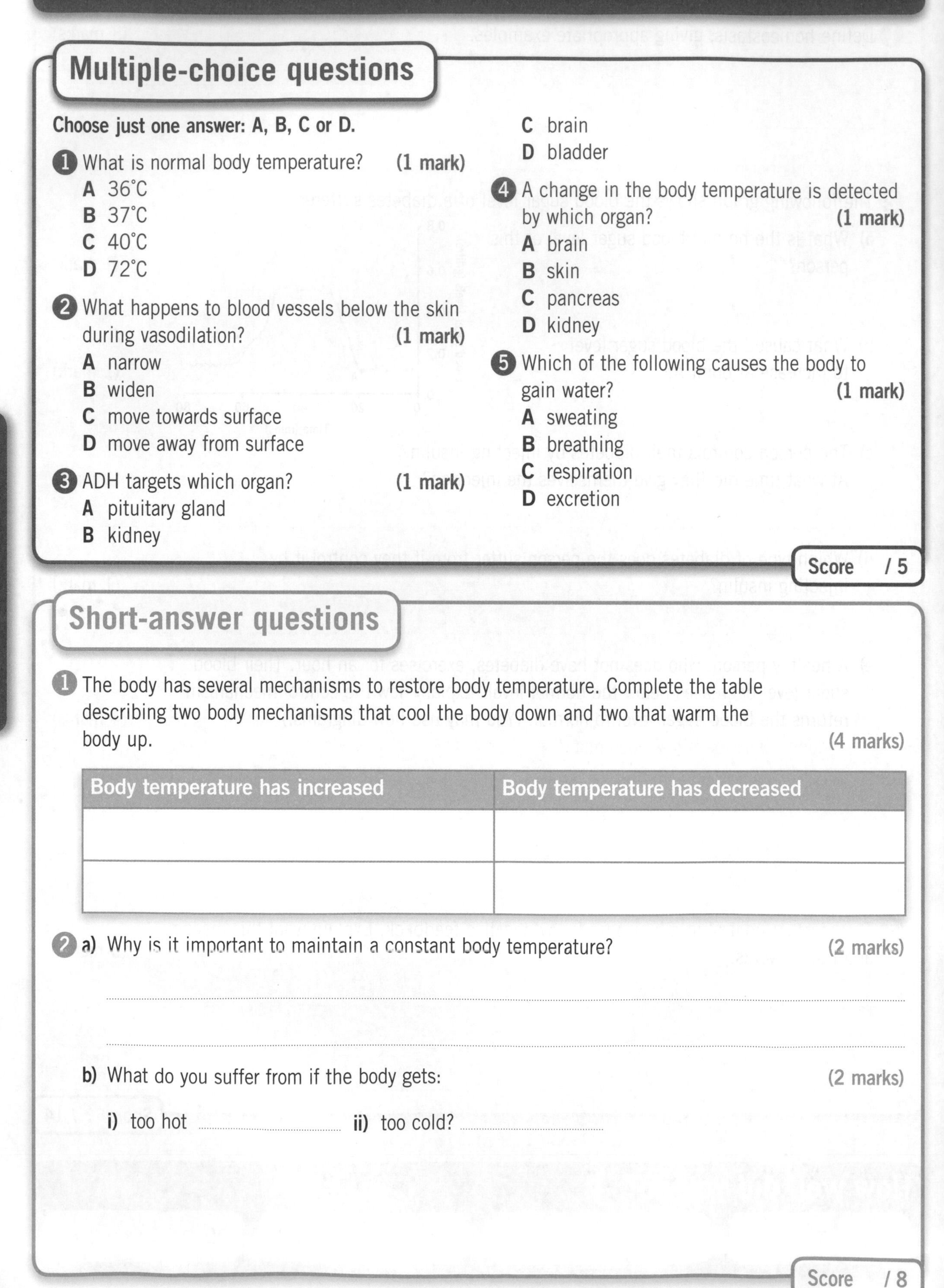

Multiple-choice questions

Choose just one answer: A, B, C or D.

1. What is normal body temperature? **(1 mark)**
   - **A** 36°C
   - **B** 37°C
   - **C** 40°C
   - **D** 72°C

2. What happens to blood vessels below the skin during vasodilation? **(1 mark)**
   - **A** narrow
   - **B** widen
   - **C** move towards surface
   - **D** move away from surface

3. ADH targets which organ? **(1 mark)**
   - **A** pituitary gland
   - **B** kidney
   - **C** brain
   - **D** bladder

4. A change in the body temperature is detected by which organ? **(1 mark)**
   - **A** brain
   - **B** skin
   - **C** pancreas
   - **D** kidney

5. Which of the following causes the body to gain water? **(1 mark)**
   - **A** sweating
   - **B** breathing
   - **C** respiration
   - **D** excretion

Score / 5

Short-answer questions

1. The body has several mechanisms to restore body temperature. Complete the table describing two body mechanisms that cool the body down and two that warm the body up. (4 marks)

Body temperature has increased	Body temperature has decreased

2. **a)** Why is it important to maintain a constant body temperature? (2 marks)

   ....

   ....

   b) What do you suffer from if the body gets: (2 marks)

   i) too hot .... **ii)** too cold? ....

Score / 8

GCSE-style questions

Answer all parts of all questions. Continue on a separate sheet of paper if necessary.

1. Complete the following sentences about water regulation. **(6 marks)**

   Osmoregulation is the control of water content in the blood. .... detect a change in the blood concentration and the hormone .... is released from the .... gland. This hormone targets the ...., which restores the water balance. The remaining waste is stored in the bladder as .... until it is ....

2. Circle the correct word. **(2 marks)**

   If a person's blood is too concentrated, **more / less** hormone is released to return the blood water content to normal. This results in the urine being **more / less** concentrated.

3. Explain how sweating helps to cool the body down. **(2 marks)**

   ....

   ....

4. **a)** Alex takes part in sports day. It is a hot day in June and he has not had time to drink all afternoon. Describe his urine that evening and which organ is involved in regulating the water content of the body. **(3 marks)**

   ....

   ....

   ....

   b) During the races Alex was hot. His body temperature control mechanisms would have been working. Describe his appearance due to these. **(2 marks)**

   ....

   ....

Score / 15

How well did you do?

0–7	Try again	8–14	Getting there	15–21	Good work	22–29	Excellent!

For more information on this topic, see pages 8–9 of your Success Revision Guide.

Hormones and Reproduction

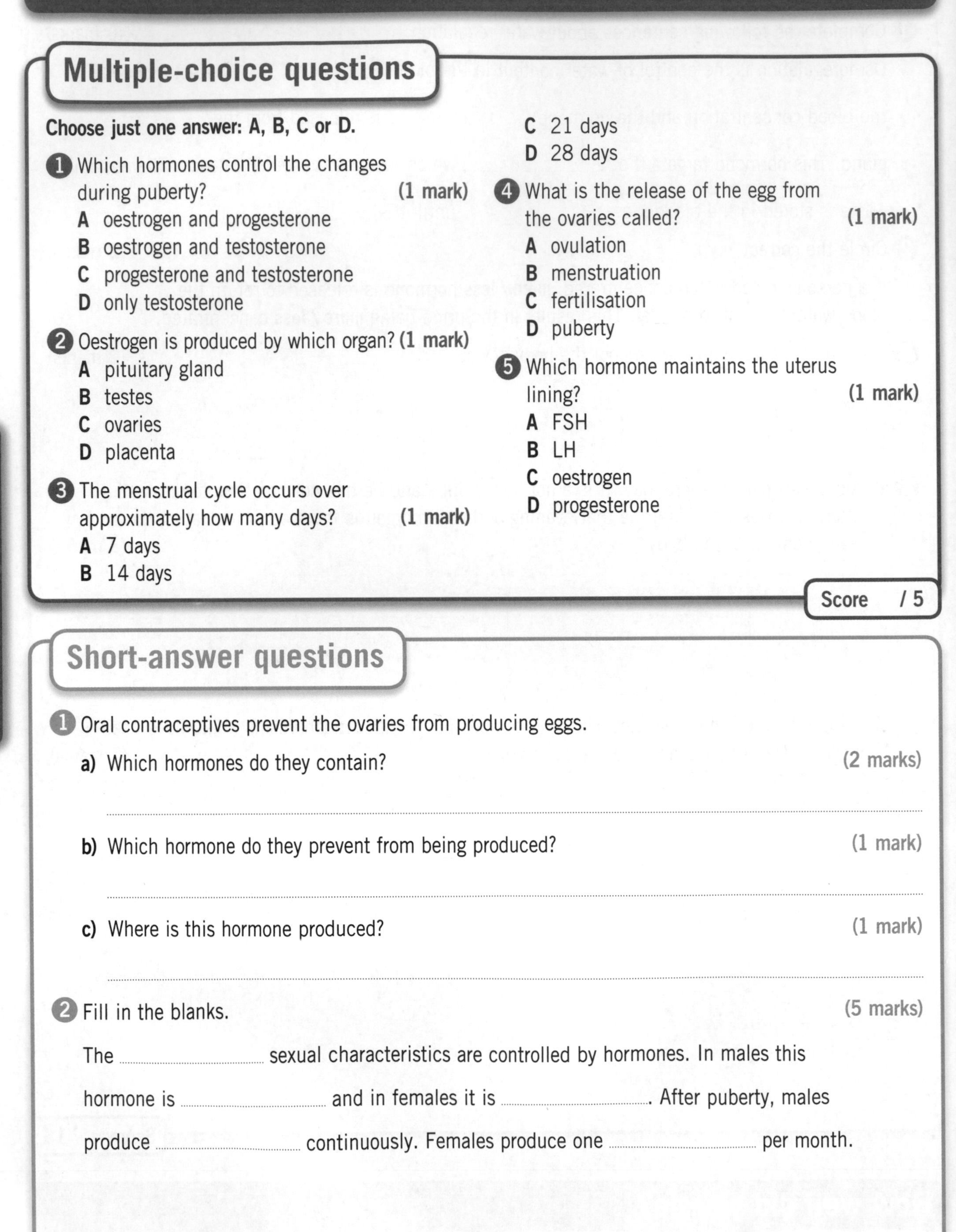

Multiple-choice questions

Choose just one answer: A, B, C or D.

1. Which hormones control the changes during puberty? **(1 mark)**
   - **A** oestrogen and progesterone
   - **B** oestrogen and testosterone
   - **C** progesterone and testosterone
   - **D** only testosterone
2. Oestrogen is produced by which organ? **(1 mark)**
   - **A** pituitary gland
   - **B** testes
   - **C** ovaries
   - **D** placenta
3. The menstrual cycle occurs over approximately how many days? **(1 mark)**
   - **A** 7 days
   - **B** 14 days
   - **C** 21 days
   - **D** 28 days
4. What is the release of the egg from the ovaries called? **(1 mark)**
   - **A** ovulation
   - **B** menstruation
   - **C** fertilisation
   - **D** puberty
5. Which hormone maintains the uterus lining? **(1 mark)**
   - **A** FSH
   - **B** LH
   - **C** oestrogen
   - **D** progesterone

Score /5

Short-answer questions

1. Oral contraceptives prevent the ovaries from producing eggs.

   a) Which hormones do they contain? (2 marks)

   ....

   b) Which hormone do they prevent from being produced? (1 mark)

   ....

   c) Where is this hormone produced? (1 mark)

   ....

2. Fill in the blanks. (5 marks)

   The .... sexual characteristics are controlled by hormones. In males this hormone is .... and in females it is .... After puberty, males produce .... continuously. Females produce one .... per month.

Score /9

GCSE-style questions

Answer all parts of all questions. Continue on a separate sheet of paper if necessary.

1. After puberty, males and females produce sex cells. Complete the table below identifying three other secondary sexual characteristics that occur for each. **(6 marks)**

Male	Female

2. The diagram shows the changes that occur in the uterus during the menstrual cycle.

a) For how many days does menstruation occur? **(1 mark)**

....

b) Which hormone repairs the uterus lining? **(1 mark)**

....

c) Which hormone triggers ovulation? .... **(1 mark)**

d) Which two hormones are produced by the ovaries and target the uterus? **(2 marks)**

....

e) At the end of the cycle the uterus lining did not break down.

i) Explain why. **(1 mark)**

....

ii) Which hormone is therefore still being produced and where is it produced? **(2 marks)**

....

3. Explain why a woman may be infertile and how she may be treated. **(6 marks)**

....

....

....

....

....

Score / 20

How well did you do?

0–8 Try again | 9–16 Getting there | 17–25 Good work | 26–34 Excellent!

For more information on this topic, see pages 10–11 of your Success Revision Guide.

Responding to the Environment

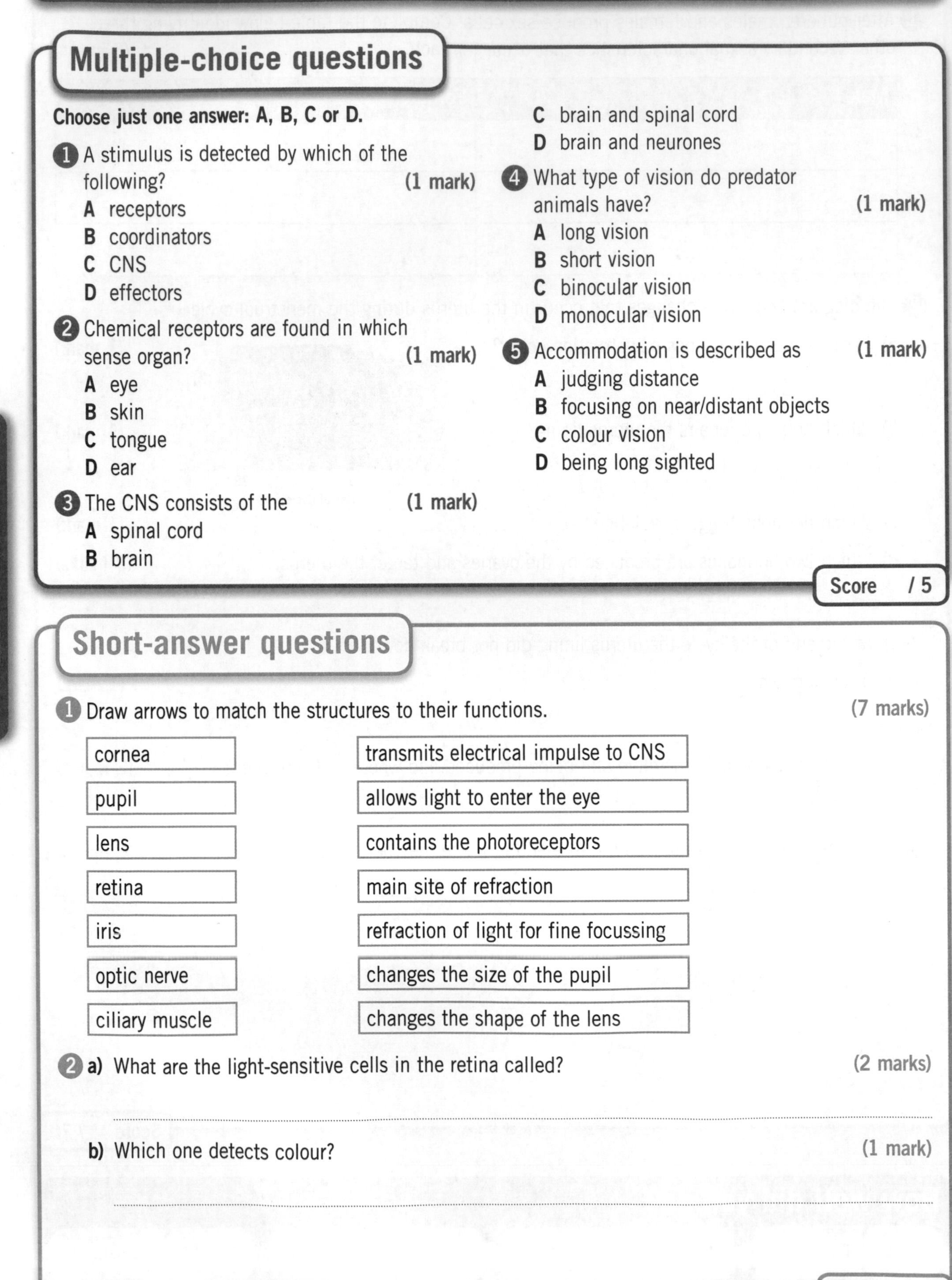

Multiple-choice questions

Choose just one answer: A, B, C or D.

1. A stimulus is detected by which of the following? **(1 mark)**
   - **A** receptors
   - **B** coordinators
   - **C** CNS
   - **D** effectors
2. Chemical receptors are found in which sense organ? **(1 mark)**
   - **A** eye
   - **B** skin
   - **C** tongue
   - **D** ear
3. The CNS consists of the **(1 mark)**
   - **A** spinal cord
   - **B** brain
   - **C** brain and spinal cord
   - **D** brain and neurones
4. What type of vision do predator animals have? **(1 mark)**
   - **A** long vision
   - **B** short vision
   - **C** binocular vision
   - **D** monocular vision
5. Accommodation is described as **(1 mark)**
   - **A** judging distance
   - **B** focusing on near/distant objects
   - **C** colour vision
   - **D** being long sighted

Score / 5

Short-answer questions

1. Draw arrows to match the structures to their functions. **(7 marks)**

Structure	Function
cornea	transmits electrical impulse to CNS
pupil	allows light to enter the eye
lens	contains the photoreceptors
retina	main site of refraction
iris	refraction of light for fine focussing
optic nerve	changes the size of the pupil
ciliary muscle	changes the shape of the lens

2. **a)** What are the light-sensitive cells in the retina called? **(2 marks)**

   ....

   b) Which one detects colour? **(1 mark)**

   ....

Score / 10

GCSE-style questions

Answer all parts of all questions. Continue on a separate sheet of paper if necessary.

1. On this diagram of an eye label structures A to D. **(4 marks)**

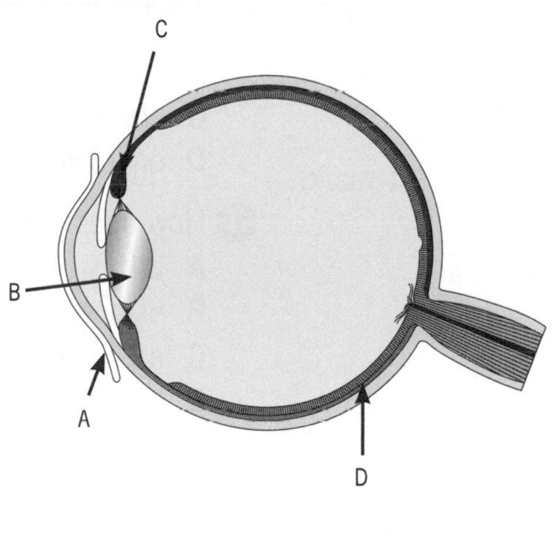

A = ... **B** = ...

C = ... **D** = ...

2. Complete the following table by crossing out the incorrect option **(4 marks)**

Object	Ciliary muscle	Suspensory ligament	Lens shape	Light refracted
Distant	**contracted / relaxed**	**tight / slack**	**fatter / thinner**	**a little / a lot**

3. Explain how organisms respond to changes in the environment. Starting with a stimulus, describe the three main stages that bring about a response. **(6 marks)**

....

....

....

....

....

....

Score / 14

How well did you do?

0–7 Try again | 8–14 Getting there | 15–21 Good work | 22–29 Excellent!

Organisms in Action

For more information on this topic, see pages 12–13 of your Success Revision Guide.

The Nervous System

Multiple-choice questions

Choose just one answer: A, B, C or D.

1 What is a synapse? **(1 mark)**

A a gap between two neurones
B the point where two neurones join
C a chemical
D a neurotransmitter

2 Which neurone carries the electrical impulse to the CNS? **(1 mark)**

A motor
B sensory
C relay
D receptor

3 Complete the following: All reflexes **(1 mark)**

A involve the brain
B are conscious responses
C are very fast
D only involve one neurone

4 How does the neurotransmitter move? **(1 mark)**

A diffusion
B osmosis
C active transport
D electrical impulses

5 Painkillers work by targeting which of the following? **(1 mark)**

A motor neurones
B sensory neurones
C relay neurones
D synapses

Score / 5

Short-answer questions

1 **a)** Which type of neurone is shown in the diagram? (1 mark)

....

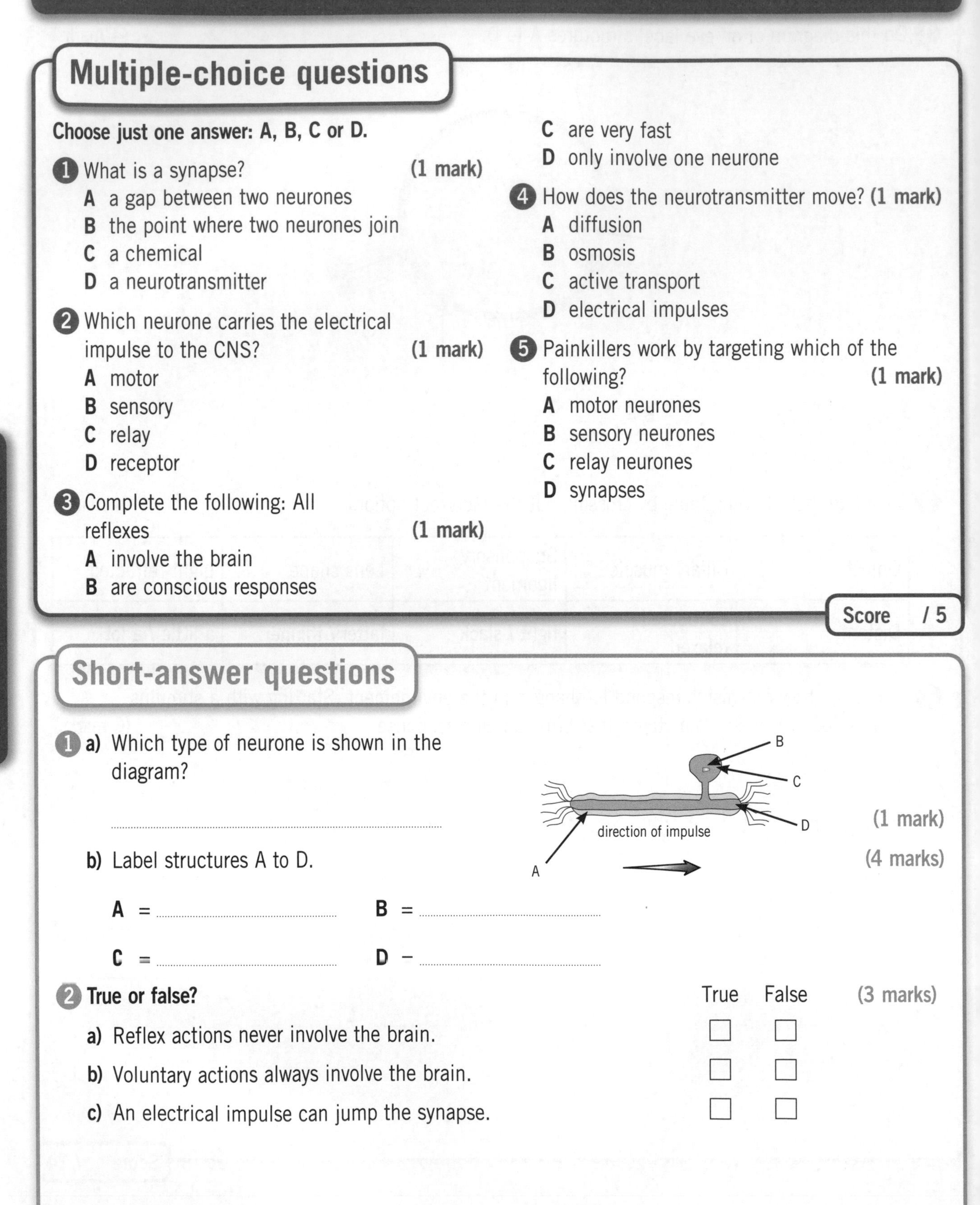

b) Label structures A to D. (4 marks)

A = ... **B** = ...

C = ... **D** – ....

2 **True or false?** (3 marks)

	True	False
a) Reflex actions never involve the brain.	☐	☐
b) Voluntary actions always involve the brain.	☐	☐
c) An electrical impulse can jump the synapse.	☐	☐

Score / 8

GCSE-style questions

Answer all parts of all questions. Continue on a separate sheet of paper if necessary.

1 Neurones are specialised cells. Describe the role of the axon, myelin sheath and dendrites. **(3 marks)**

....

....

....

2 These sentences describe how the impulse crosses the synapse, but they are in the wrong order. Fill in the boxes below to show the correct order. **(4 marks)**

A Neurotransmitter diffuses across the gap.

B Neurotransmitter joins to receptors.

C Neurotransmitter is released from sensory neurone.

D Electrical impulse formed in relay neurone.

3 The diagram shows the events that occur when a person touches a pin.

Describe the reflex action. **(6 marks)**

....

....

....

....

....

Score / 13

How well did you do?

0–6 Try again | 7–13 Getting there | 14–19 Good work | 20–26 Excellent!

For more information on this topic, see pages 14–15 of your Success Revision Guide.

Plant Responses

Multiple-choice questions

Choose just one answer: A, B, C or D.

1 Which word describes plant growth, in a particular direction, due to external stimuli? **(1 mark)**

A coordination
B communication
C tropism
D elongation

2 Which of the following describes how roots grow? **(1 mark)**

A away from light and towards water
B away from light
C towards gravity
D away from light and towards gravity

3 Which of the following gardening activities does not involve the use of plant hormones? **(1 mark)**

A killing weeds
B increasing photosynthesis
C producing seedless fruit
D ripening fruit

4 Which of the following does auxin cause? **(1 mark)**

A cell elongation
B cell coordination
C cell division
D cell differentiation

5 Which stimulus negatively controls shoot growth of a seed? **(1 mark)**

A light
B water
C heat
D gravity

Score / 5

Short-answer questions

1 Plants respond to external stimuli.

a) What is the response to light called? (1 mark)

....

b) What is the response to gravity called? (1 mark)

....

c) Which hormone brings about these responses? (1 mark)

....

2 Gardeners make use of plant hormones. Give three uses of plant hormones. (3 marks)

a) ....

b) ....

c) ....

Score / 6

GCSE-style questions

Answer all parts of all questions. Continue on a separate sheet of paper if necessary.

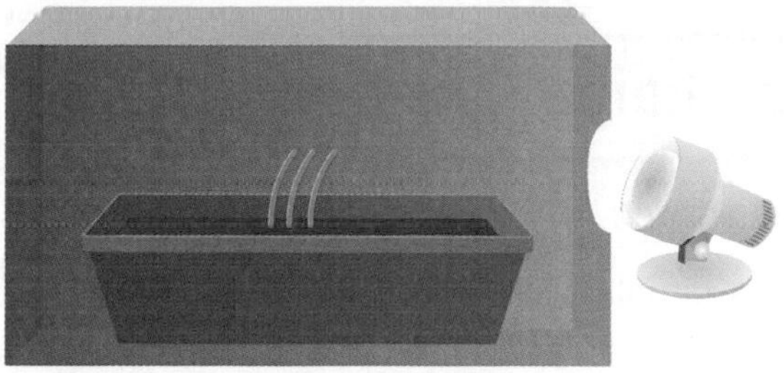

1 a) Describe the growth of the shoots in the diagram. **(1 mark)**

....

....

b) Explain what is happening within the shoots to make this occur. **(3 marks)**

....

....

....

2 The diagrams illustrate the response of a shoot and a root when a plant is placed on its side. Using the diagrams, explain what will happen to both the roots and stems. **(6 marks)**

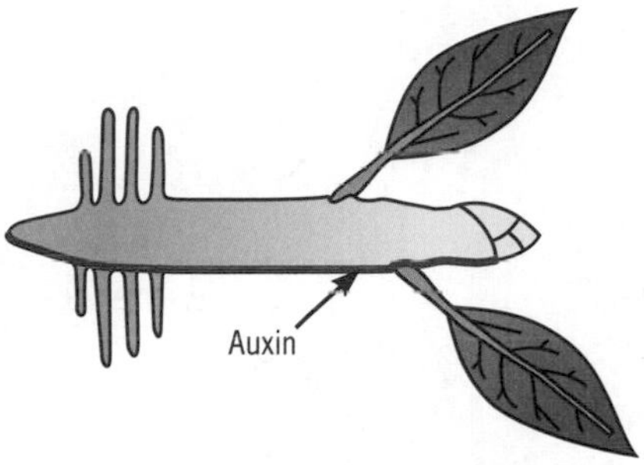

....

....

....

....

....

....

....

....

....

....

....

3 Seeds germinate underground. Explain why the shoot still grows towards the light. **(1 mark)**

....

Score / 11

How well did you do?

0–5	Try again	6–11	Getting there	12–17	Good work	18–22	Excellent!

For more information on this topic, see pages 16–17 of your Success Revision Guide.

Pathogens and Infections

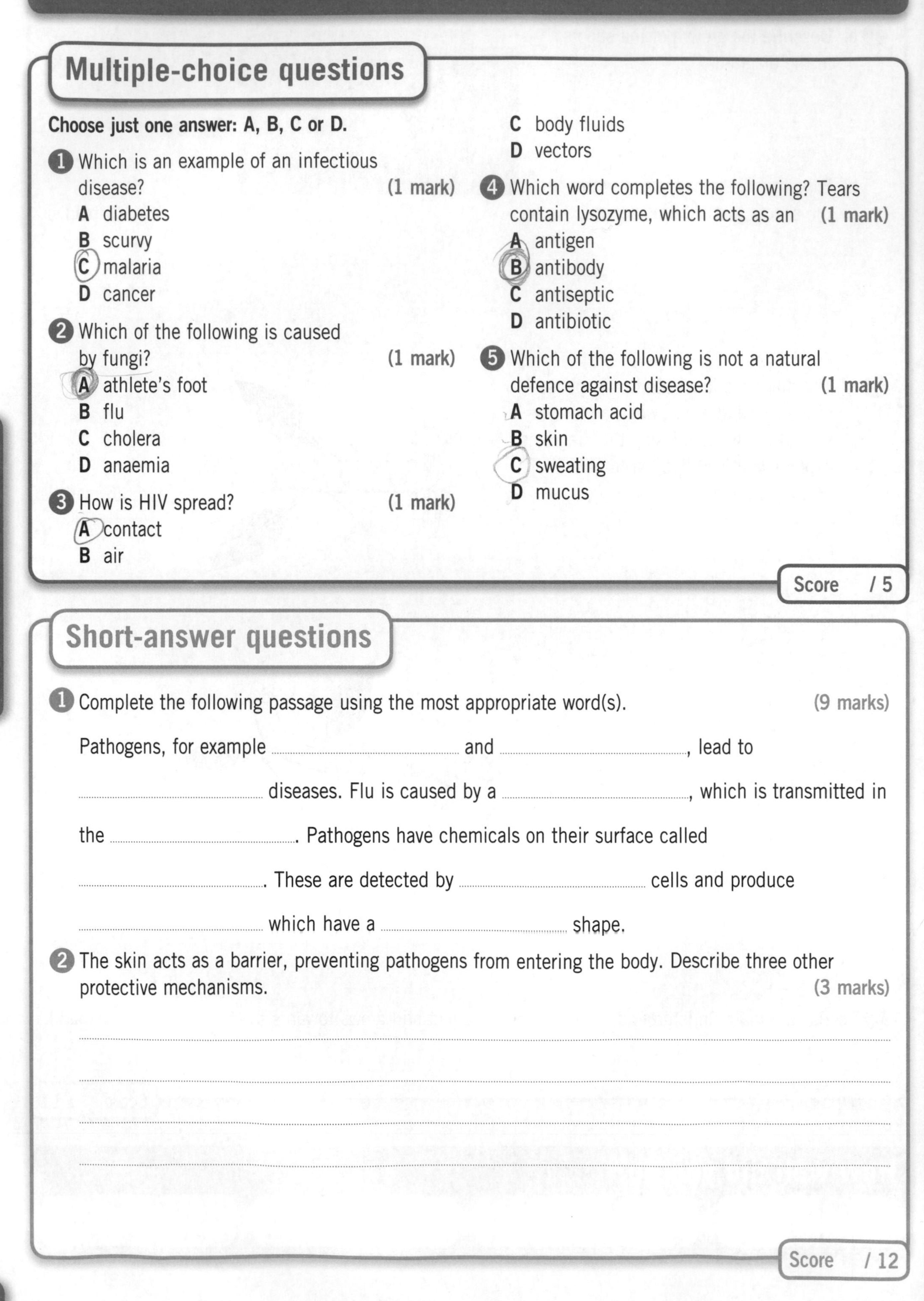

Multiple-choice questions

Choose just one answer: A, B, C or D.

1. Which is an example of an infectious disease? **(1 mark)**
   A diabetes
   B scurvy
   C malaria
   D cancer

2. Which of the following is caused by fungi? **(1 mark)**
   A athlete's foot
   B flu
   C cholera
   D anaemia

3. How is HIV spread? **(1 mark)**
   A contact
   B air
   C body fluids
   D vectors

4. Which word completes the following? Tears contain lysozyme, which acts as an **(1 mark)**
   A antigen
   B antibody
   C antiseptic
   D antibiotic

5. Which of the following is not a natural defence against disease? **(1 mark)**
   A stomach acid
   B skin
   C sweating
   D mucus

Score /5

Short-answer questions

1. Complete the following passage using the most appropriate word(s). **(9 marks)**

   Pathogens, for example .... and .... , lead to .... diseases. Flu is caused by a .... , which is transmitted in the .... Pathogens have chemicals on their surface called .... These are detected by .... cells and produce .... which have a .... shape.

2. The skin acts as a barrier, preventing pathogens from entering the body. Describe three other protective mechanisms. **(3 marks)**

   ....

   ....

   ....

   ....

Score /12

GCSE-style questions

Answer all parts of all questions. Continue on a separate sheet of paper if necessary.

1. Malaria is an infectious disease.

   a) Which type of organism is it caused by? Mosquitos. **(1 mark)**

   b) How is it transmitted? **(2 marks)**

   c) How can its spread be controlled? **(3 marks)**

2. Distinguish between the following.

   a) antigens and antibodies **(2 marks)**

   b) lymphocyte and phagocyte **(2 marks)**

3. True or false? **(4 marks)**

	True	False
a) Flu is transmitted by body fluids.	☐	☐
b) Red-green colour blindness is a deficiency disease.	☐	☐
c) Stomach acid is useful in preventing salmonella entering the body.	☐	☐
d) Memory cells are produced by the brain.	☐	☐

4. **a)** Explain how phagocytes kill bacteria. **(2 marks)**

   b) A person has had chicken pox. Explain why this person will not suffer from chicken pox again. **(4 marks)**

Score / 20

How well did you do?

0–9	Try again	10–18	Getting there	19–27	Good work	28–37	Excellent!

For more information on this topic, see pages 20–21 of your Success Revision Guide.

Antibiotics and Antiseptics

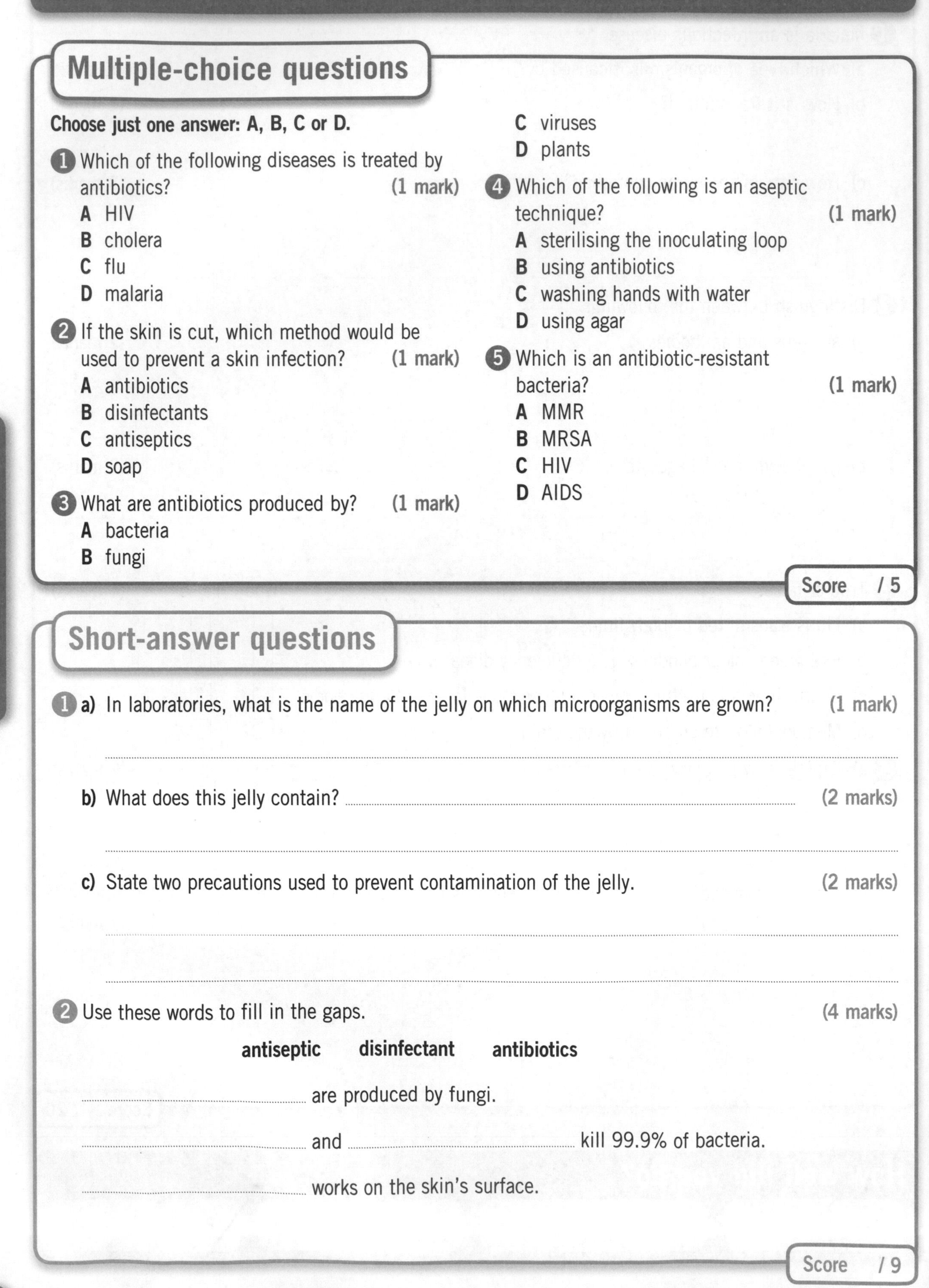

Multiple-choice questions

Choose just one answer: A, B, C or D.

1. Which of the following diseases is treated by antibiotics? **(1 mark)**
   A HIV
   B cholera
   C flu
   D malaria

2. If the skin is cut, which method would be used to prevent a skin infection? **(1 mark)**
   A antibiotics
   B disinfectants
   C antiseptics
   D soap

3. What are antibiotics produced by? **(1 mark)**
   A bacteria
   B fungi
   C viruses
   D plants

4. Which of the following is an aseptic technique? **(1 mark)**
   A sterilising the inoculating loop
   B using antibiotics
   C washing hands with water
   D using agar

5. Which is an antibiotic-resistant bacteria? **(1 mark)**
   A MMR
   B MRSA
   C HIV
   D AIDS

Score /5

Short-answer questions

1. **a)** In laboratories, what is the name of the jelly on which microorganisms are grown? **(1 mark)**

   ....

   b) What does this jelly contain? .... **(2 marks)**

   ....

   c) State two precautions used to prevent contamination of the jelly. **(2 marks)**

   ....

   ....

2. Use these words to fill in the gaps. **(4 marks)**

   antiseptic **disinfectant** **antibiotics**

   .... are produced by fungi.

   .... and .... kill 99.9% of bacteria.

   .... works on the skin's surface.

Score /9

Health and Disease

GCSE-style questions

Answer all parts of all questions. Continue on a separate sheet of paper if necessary.

1 An investigation into the effectiveness of antibiotics was carried out. Discs containing antibiotic A to H were placed on agar that had bacteria growing on it. The diagram below shows the results.

a) Why are the bacteria grown on agar? **(1 mark)**

....

....

b) Explain why some discs did not kill the bacteria. **(1 mark)**

....

c) Which was the best antibiotic for this bacteria? Explain your reason. **(2 marks)**

....

....

d) Name a common antibiotic. **(1 mark)**

....

2 Antibiotic-resistant bacteria are becoming an increasing problem for doctors.

a) Explain how bacteria become resistant. .... **(1 mark)**

....

b) What precautions should we take to prevent more bacteria becoming resistant? **(2 marks)**

....

....

c) How are hospitals trying to prevent the spread of antibiotic resistance? **(3 marks)**

....

....

....

Score /11

How well did you do?

0–6	Try again	7–12	Getting there	13–19	Good work	20–25	Excellent!

For more information on this topic, see pages 22–23 of your Success Revision Guide.

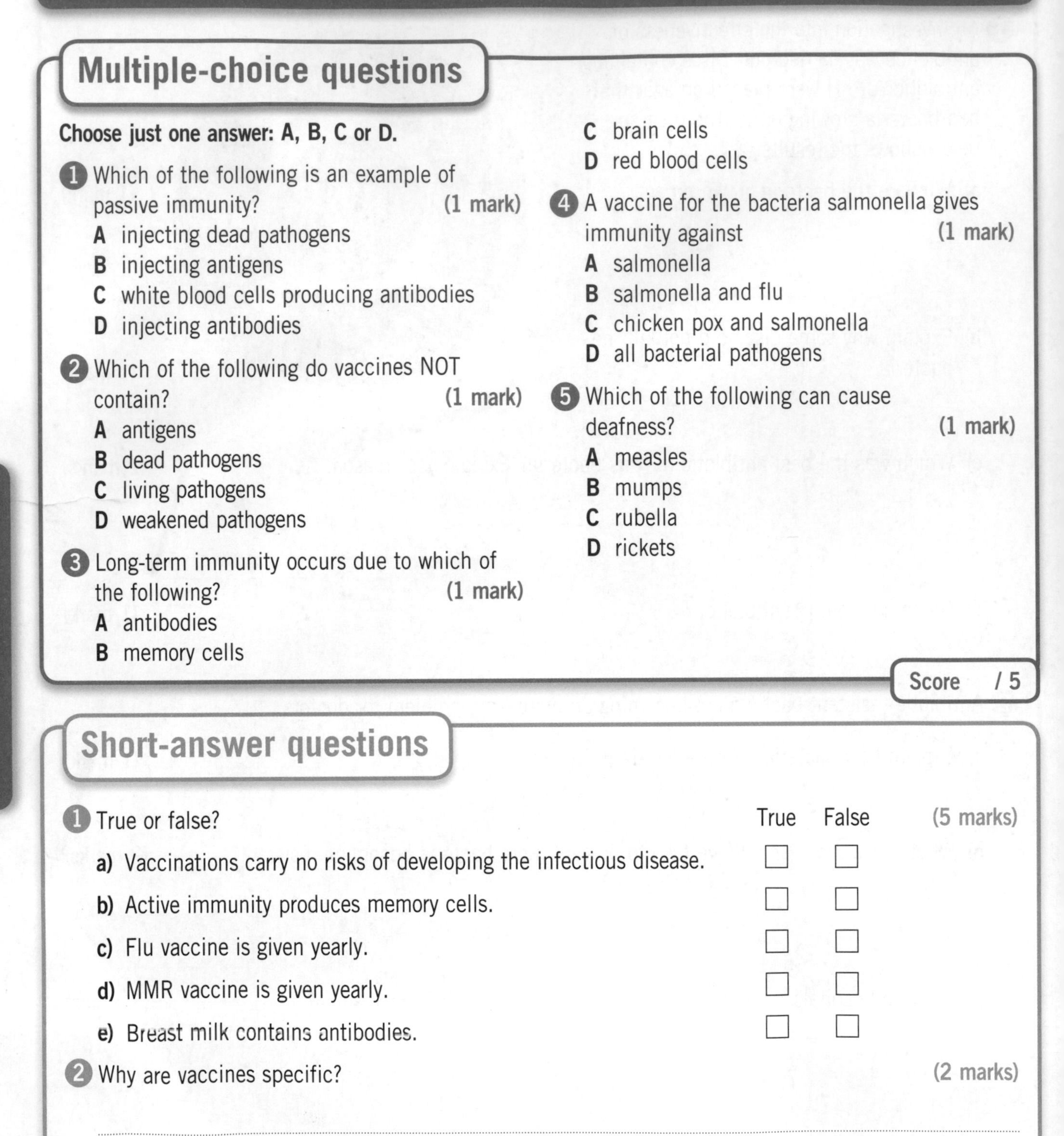

Vaccinations

Multiple-choice questions

Choose just one answer: A, B, C or D.

1. Which of the following is an example of passive immunity? **(1 mark)**
   A injecting dead pathogens
   B injecting antigens
   C white blood cells producing antibodies
   D injecting antibodies

2. Which of the following do vaccines NOT contain? **(1 mark)**
   A antigens
   B dead pathogens
   C living pathogens
   D weakened pathogens

3. Long-term immunity occurs due to which of the following? **(1 mark)**
   A antibodies
   B memory cells
   C brain cells
   D red blood cells

4. A vaccine for the bacteria salmonella gives immunity against **(1 mark)**
   A salmonella
   B salmonella and flu
   C chicken pox and salmonella
   D all bacterial pathogens

5. Which of the following can cause deafness? **(1 mark)**
   A measles
   B mumps
   C rubella
   D rickets

Score / 5

Short-answer questions

1. True or false? **(5 marks)**

	True	False
a) Vaccinations carry no risks of developing the infectious disease.	☐	☐
b) Active immunity produces memory cells.	☐	☐
c) Flu vaccine is given yearly.	☐	☐
d) MMR vaccine is given yearly.	☐	☐
e) Breast milk contains antibodies.	☐	☐

2. Why are vaccines specific? **(2 marks)**

....

....

Score / 7

GCSE-style questions

Answer all parts of all questions. Continue on a separate sheet of paper if necessary.

1 **a)** Why may someone be injected with an antibody? **(1 mark)**

....

b) How does this immunity differ from when the antibodies are produced by the person's white blood cells? **(2 marks)**

....

....

2 ✎ Distinguish between active and passive immunity, giving examples of each. **(6 marks)**

....

....

....

....

....

3 **a)** Which diseases does MMR protect against? **(3 marks)**

....

b) What is the risk to a pregnant woman if she has not had her MMR? **(1 mark)**

....

c) Suggest two concerns regarding giving the MMR vaccination. **(2 marks)**

....

....

....

4 Explain why the flu vaccine is only effective for one year **(2 marks)**

....

....

Score / 17

How well did you do?

0–7	Try again	8–14	Getting there	15–21	Good work	22–29	Excellent!

For more information on this topic, see pages 24–25 of your Success Revision Guide.

Drugs

Multiple-choice questions

Choose just one answer: A, B, C or D.

1. Alcohol is a (1 mark)
   - **A** depressant
   - **B** stimulant
   - **C** placebo
   - **D** painkiller

2. Which class of drugs is the most dangerous? (1 mark)
   - **A** A
   - **B** B
   - **C** C
   - **D** X

3. Which of the following is a class A drug? (1 mark)
   - **A** nicotine
   - **B** antibiotics
   - **C** cannabis
   - **D** heroin

4. Which type of drug blocks the transmission of nerve impulses? (1 mark)
   - **A** depressants
   - **B** stimulants
   - **C** placebos
   - **D** class A

5. Which of the following best describes cannabis? (1 mark)
   - **A** class A
   - **B** available on prescription
   - **C** illegal
   - **D** not addictive

Score /5

Short-answer questions

1. Define the following terms.

   a) Addictive .... (1 mark)

   b) Tolerance .... (1 mark)

2. **a)** Thalidomide was used to treat morning sickness in pregnant woman. Why was the drug later banned from use in pregnancy? (1 mark)

   ....

   b) What is thalidomide now used to treat? .... (1 mark)

3. True or false? (3 marks)

	True	False
a) Cocaine is a class C drug.	☐	☐
b) Drugs alter the functioning of the body.	☐	☐
c) Double blind testing of drugs is when only the doctor knows which the placebo is.	☐	☐

Score /7

GCSE-style questions

Answer all parts of all questions. Continue on a separate sheet of paper if necessary.

1 A new drug is being developed to treat a skin condition.

a) Why does the drug need to be tested before it can be used on patients? **(2 marks)**

....

....

b) How will the drug be tested? **(2 marks)**

....

....

c) What are the ethical issues of testing on animals? **(2 marks)**

....

....

2 Drug A was trialled on ten patients. The patients were split into two groups. One group was given drug A; the other group was given a placebo. It was a blind test.

a) What is a placebo and why is it given? **(2 marks)**

....

....

b) What does 'it was a blind test' mean? **(1 mark)**

....

c) What is the advantage of carrying out a 'blind test'? **(1 mark)**

....

d) How could this trial be made more reliable? **(1 mark)**

....

3 Compare the effects of a stimulant and a depressant on the nervous system. **(2 marks)**

....

....

....

Score / 13

How well did you do?

0–6	Try again	7–12	Getting there	13–19	Good work	20–25	Excellent!

For more information on this topic, see pages 26–27 of your Success Revision Guide.

Smoking and Drinking

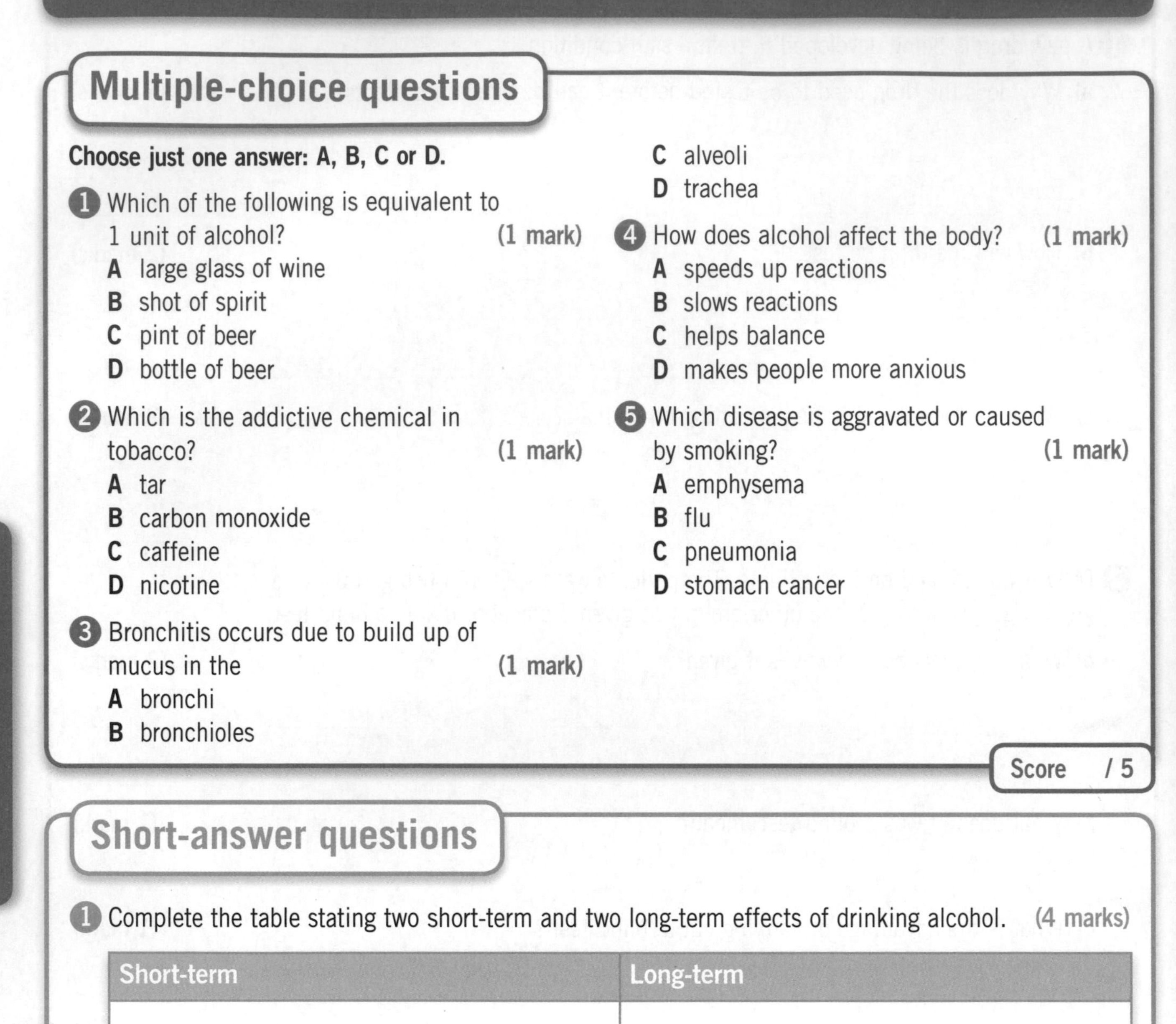

Multiple-choice questions

Choose just one answer: A, B, C or D.

1. Which of the following is equivalent to 1 unit of alcohol? **(1 mark)**
   A large glass of wine
   B shot of spirit
   C pint of beer
   D bottle of beer

2. Which is the addictive chemical in tobacco? **(1 mark)**
   A tar
   B carbon monoxide
   C caffeine
   D nicotine

3. Bronchitis occurs due to build up of mucus in the **(1 mark)**
   A bronchi
   B bronchioles
   C alveoli
   D trachea

4. How does alcohol affect the body? **(1 mark)**
   A speeds up reactions
   B slows reactions
   C helps balance
   D makes people more anxious

5. Which disease is aggravated or caused by smoking? **(1 mark)**
   A emphysema
   B flu
   C pneumonia
   D stomach cancer

Score / 5

Short-answer questions

1. Complete the table stating two short-term and two long-term effects of drinking alcohol. **(4 marks)**

Short-term	Long-term

2. Why is alcohol described in units? **(1 mark)**

   ....

3. State three diseases aggravated or caused by smoking. **(3 marks)**

   ....

   ....

   ....

Score / 8

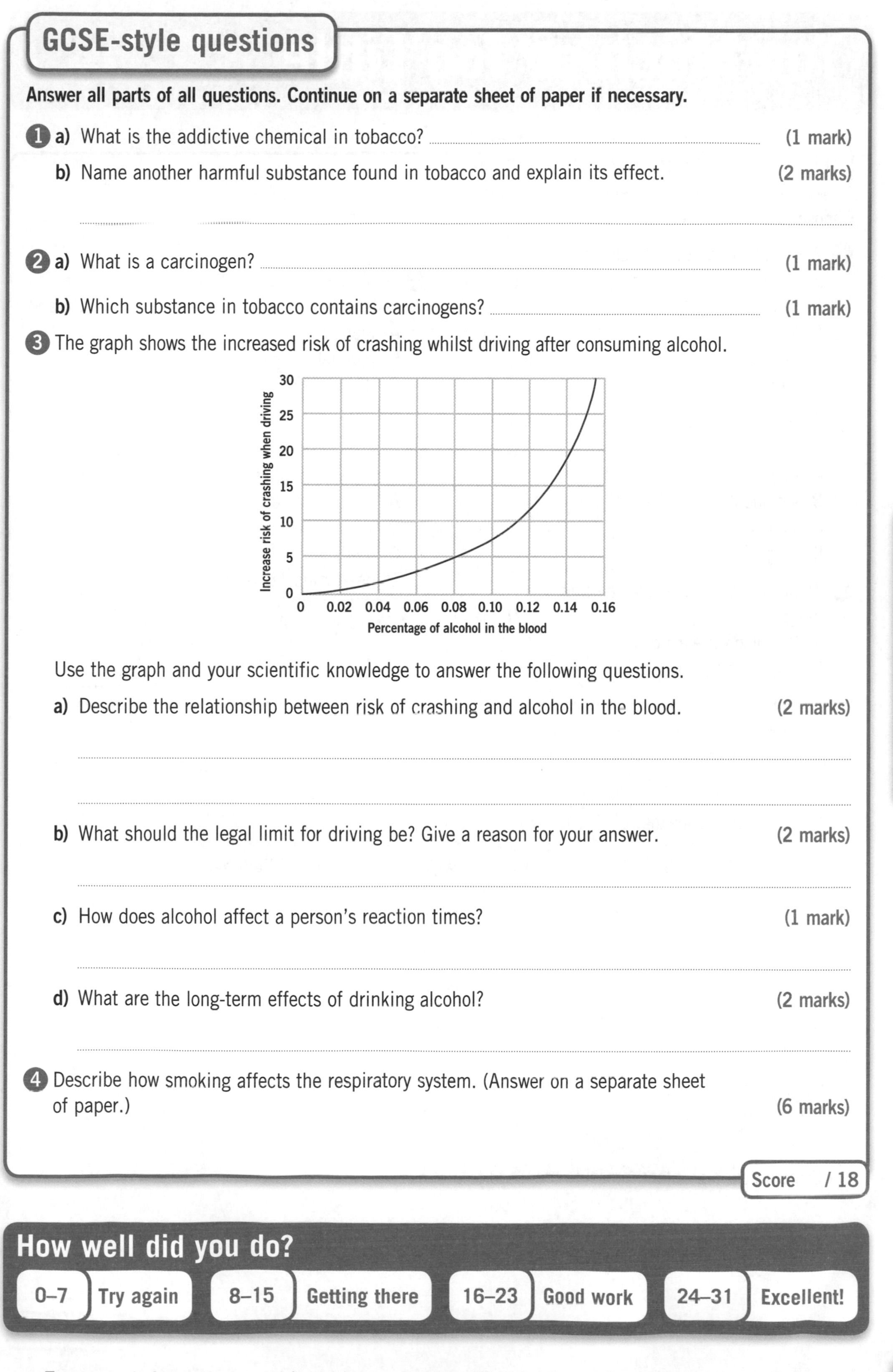

GCSE-style questions

Answer all parts of all questions. Continue on a separate sheet of paper if necessary.

1 **a)** What is the addictive chemical in tobacco? .... **(1 mark)**

b) Name another harmful substance found in tobacco and explain its effect. **(2 marks)**

....

2 **a)** What is a carcinogen? .... **(1 mark)**

b) Which substance in tobacco contains carcinogens? .... **(1 mark)**

3 The graph shows the increased risk of crashing whilst driving after consuming alcohol.

Use the graph and your scientific knowledge to answer the following questions.

a) Describe the relationship between risk of crashing and alcohol in the blood. **(2 marks)**

....

....

b) What should the legal limit for driving be? Give a reason for your answer. **(2 marks)**

....

c) How does alcohol affect a person's reaction times? **(1 mark)**

....

d) What are the long-term effects of drinking alcohol? **(2 marks)**

....

4 Describe how smoking affects the respiratory system. (Answer on a separate sheet of paper.) **(6 marks)**

Score / 18

How well did you do?

0–7	Try again	8–15	Getting there	16–23	Good work	24–31	Excellent!

For more information on this topic, see pages 28–29 of your Success Revision Guide.

Too Much or Too Little

Multiple-choice questions

Choose just one answer: A, B, C or D.

1. Which of the following helps reduce blood pressure? **(1 mark)**
   - **A** salt
   - **B** alcohol
   - **C** exercise
   - **D** stress
2. Obesity has been linked to which form of cancer? **(1 mark)**
   - **A** lung
   - **B** stomach
   - **C** skin
   - **D** breast
3. What does dizziness and fainting indicate? **(1 mark)**
   - **A** low blood pressure
   - **B** high blood pressure
   - **C** obesity
   - **D** thrombosis
4. What is a fatty deposit inside a blood vessel called? **(1 mark)**
   - **A** plaque
   - **B** platelets
   - **C** thrombosis
   - **D** thrombin
5. What is systolic pressure a measure of? The pressure when **(1 mark)**
   - **A** arteries contract
   - **B** veins relax
   - **C** heart muscles contract
   - **D** heart muscles relax

Score / 5

Short-answer questions

1. Complete the following. **(4 marks)**

   The heart is made of .... cells, which need a constant supply of .... and .... Fatty deposits form in the blood vessels supplying the heart. These blood vessels are called .... This may lead to a heart attack.

2. List four factors which increase a person's chance of heart disease. **(4 marks)**

   ....

   ....

   ....

3. Explain how high blood pressure may lead to a stroke. **(2 marks)**

   ....

   ....

Score / 10

GCSE-style questions

Answer all parts of all questions. Continue on a separate sheet of paper if necessary.

1

$$\text{Body Mass Index (BMI)} = \frac{\text{mass in kg}}{(\text{height in metres})^2}$$

a) Calculate the BMI of a person who has a mass of 90 kg and height of 164 cm. Show your working. **(2 marks)**

b) A BMI of above 30 indicates that a person is obese. What health risks are associated with obesity? **(3 marks)**

....

....

c) How could an obese person reduce their BMI? **(2 marks)**

....

2 True or false? **(2 marks)**

	True	False
a) Obesity lowers blood pressure.	☐	☐
b) A stroke occurs when blood vessels burst in the brain.	☐	☐
c) Anaemia occurs due to a lack of vitamin C.	☐	☐

3 A person has a blood pressure of 120 over 80.

a) What does 120 mean? **(1 mark)**

....

b) What advice would you give someone with a high blood pressure in order to help them lower it? **(3 marks)**

....

....

4 Define these diseases. **(4 marks)**

diabetes ....

thrombosis ....

stroke ....

kwashiorkor ....

Score / 17

How well did you do?

0–8	Try again	9–16	Getting there	17–24	Good work	27–32	Excellent!

For more information on this topic, see pages 30–31 of your Success Revision Guide.

Genes and Chromosomes

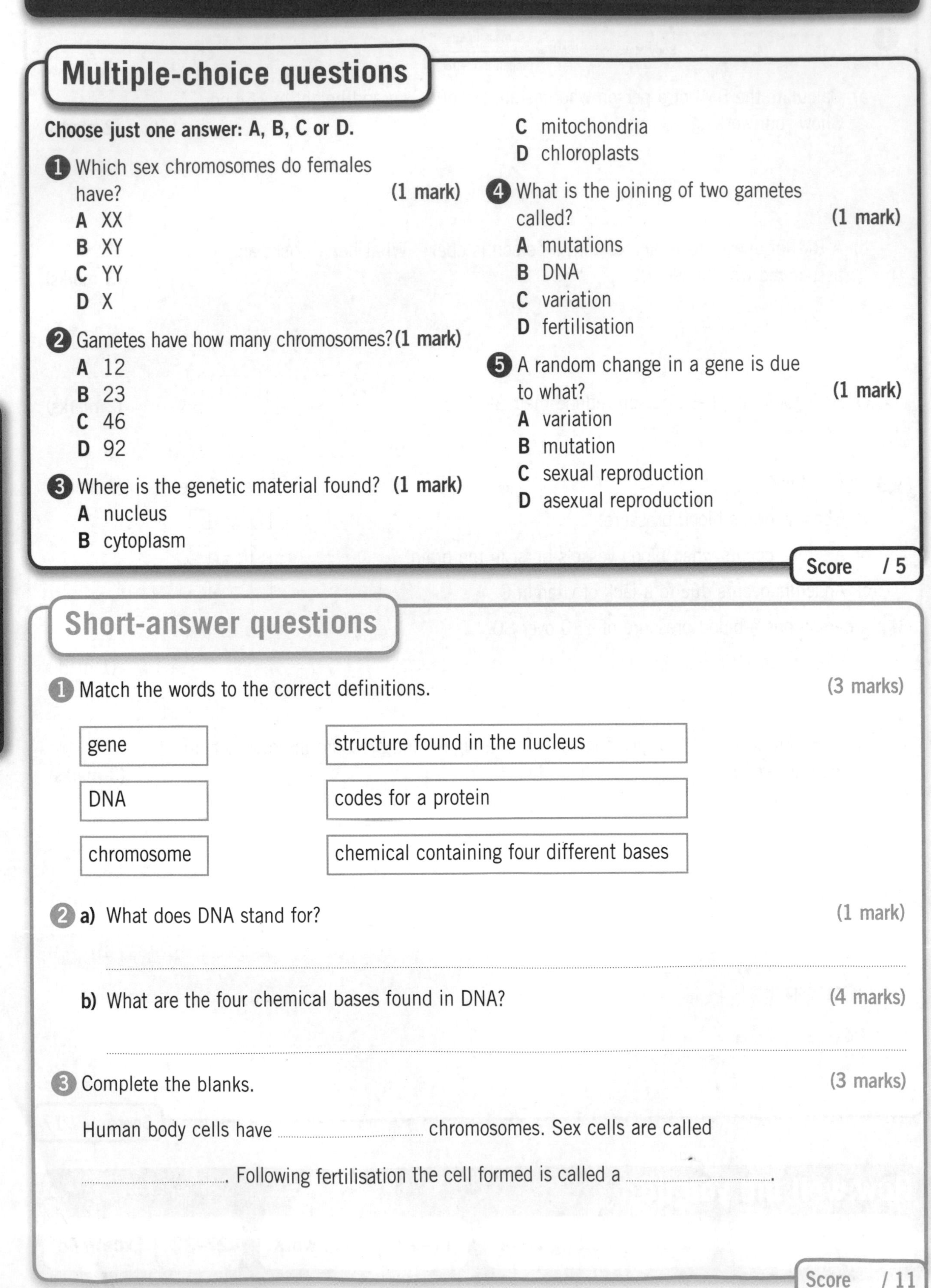

Multiple-choice questions

Choose just one answer: A, B, C or D.

1. Which sex chromosomes do females have? **(1 mark)**
   - **A** XX
   - **B** XY
   - **C** YY
   - **D** X
2. Gametes have how many chromosomes? **(1 mark)**
   - **A** 12
   - **B** 23
   - **C** 46
   - **D** 92
3. Where is the genetic material found? **(1 mark)**
   - **A** nucleus
   - **B** cytoplasm
   - **C** mitochondria
   - **D** chloroplasts
4. What is the joining of two gametes called? **(1 mark)**
   - **A** mutations
   - **B** DNA
   - **C** variation
   - **D** fertilisation
5. A random change in a gene is due to what? **(1 mark)**
   - **A** variation
   - **B** mutation
   - **C** sexual reproduction
   - **D** asexual reproduction

Score / 5

Short-answer questions

1. Match the words to the correct definitions. (3 marks)

gene	structure found in the nucleus
DNA	codes for a protein
chromosome	chemical containing four different bases

2. **a)** What does DNA stand for? (1 mark)

   ....

   b) What are the four chemical bases found in DNA? (4 marks)

   ....

3. Complete the blanks. (3 marks)

   Human body cells have .... chromosomes. Sex cells are called .... Following fertilisation the cell formed is called a ....

Score / 11

GCSE-style questions

Answer all parts of all questions. Continue on a separate sheet of paper if necessary.

1 a) Complete the diagram showing how sex is determined. **(3 marks)**

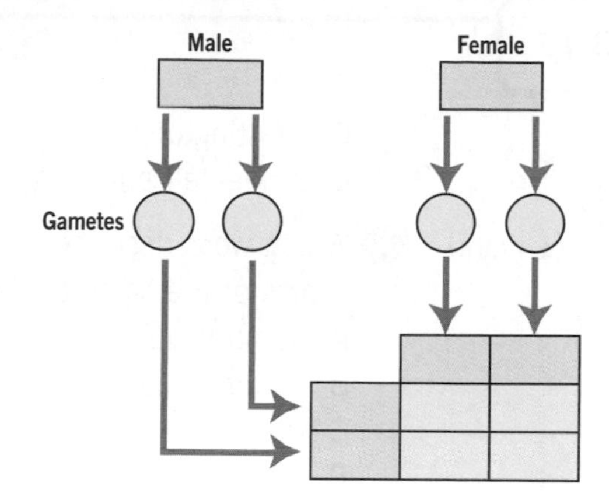

b) A couple have two daughters. What is the probability that their next child will be a daughter? **(1 mark)**

....

2 Explain why sexual reproduction results in variation. **(2 marks)**

....

....

3 Differences are due to variation. Complete the table below by

a) stating the two causes of variation **(2 marks)**

b) giving two examples of each. **(4 marks)**

Cause of variation	Examples
	1. 2.
	1. 2.

Score / 12

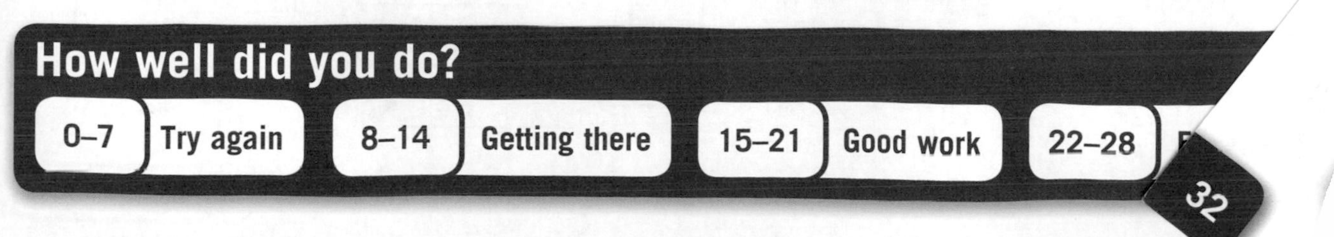

For more information on this topic, see pages 34–35 of your Success Revision Gui

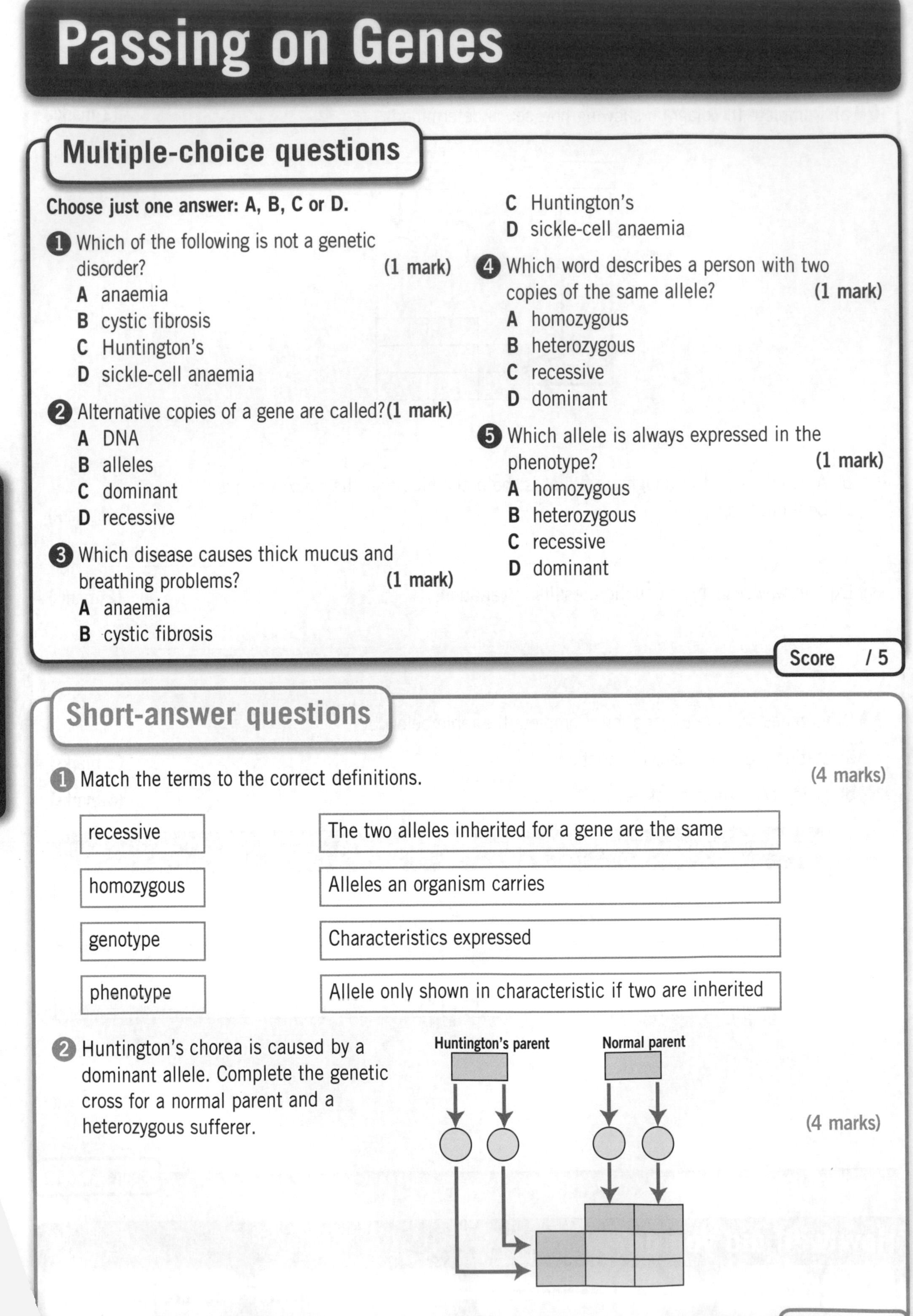

Passing on Genes

Multiple-choice questions

Choose just one answer: A, B, C or D.

1. Which of the following is not a genetic disorder? **(1 mark)**
   - **A** anaemia
   - **B** cystic fibrosis
   - **C** Huntington's
   - **D** sickle-cell anaemia
2. Alternative copies of a gene are called? **(1 mark)**
   - **A** DNA
   - **B** alleles
   - **C** dominant
   - **D** recessive
3. Which disease causes thick mucus and breathing problems? **(1 mark)**
   - **A** anaemia
   - **B** cystic fibrosis
   - **C** Huntington's
   - **D** sickle-cell anaemia
4. Which word describes a person with two copies of the same allele? **(1 mark)**
   - **A** homozygous
   - **B** heterozygous
   - **C** recessive
   - **D** dominant
5. Which allele is always expressed in the phenotype? **(1 mark)**
   - **A** homozygous
   - **B** heterozygous
   - **C** recessive
   - **D** dominant

Score /5

Short-answer questions

1. Match the terms to the correct definitions. (4 marks)

Term	Definition
recessive	The two alleles inherited for a gene are the same
homozygous	Alleles an organism carries
genotype	Characteristics expressed
phenotype	Allele only shown in characteristic if two are inherited

2. Huntington's chorea is caused by a dominant allele. Complete the genetic cross for a normal parent and a heterozygous sufferer. (4 marks)

Score /8

GCSE-style questions

Answer all parts of all questions. Continue on a separate sheet of paper if necessary.

1. Genetic screening can be carried out on a foetus.

   a) Give an advantage and a disadvantage of genetic screening for the parents. **(2 marks)**

   b) What are the ethical considerations? **(2 marks)**

2. Cystic fibrosis is a genetic disease caused by a recessive allele. A couple who are both carriers of the recessive allele for cystic fibrosis are expecting a child. What is the probability their child will suffer from cystic fibrosis? Draw a genetic diagram on a separate sheet of paper. **(5 marks)**

   Probability =

3. Sickle cell anaemia is a genetic disease that affects the red blood cells.

   a) Is it caused by a dominant or a recessive allele? **(1 mark)**

   b) What are the symptoms of the disease? **(2 marks)**

4. The diagram shows a family tree for the inheritance of cystic fibrosis.

   Fred — Maria

   John, Harry

   O = sufferer
   O = non-sufferer

   a) What evidence is there that cystic fibrosis is coded for by a recessive allele? **(2 marks)**

   b) Are cystic fibrosis sufferers heterozygous or homozygous? **(1 mark)**

   c) What are the symptoms of cystic fibrosis? **(2 marks)**

Score / 17

How well did you do?

0–7	Try again	8–15	Getting there	16–23	Good work	24–30	Excellent!

For more information on this topic, see pages 36–37 of your Success Revision Guide.

Introduction to Gene Technology

Multiple-choice questions

Choose just one answer: A, B, C or D.

1. Genetic diseases can be treated using which of the following? **(1 mark)**
   - **A** stem cells
   - **B** clones
   - **C** tissue culture
   - **D** body cells

2. What is a clone? **(1 mark)**
   - **A** similar organisms
   - **B** different organisms
   - **C** genetically identical organisms
   - **D** genetically engineered organisms

3. Dolly the sheep was produced by which method? **(1 mark)**
   - **A** cloning adult cells
   - **B** cloning embryos
   - **C** gene therapy
   - **D** genetic engineering

4. Which of the following can be treated by gene therapy? **(1 mark)**
   - **A** obesity
   - **B** heart disease
   - **C** scurvy
   - **D** cystic fibrosis

5. What is the term for growing a new plant from a small group of cells? **(1 mark)**
   - **A** genetic modification
   - **B** cuttings
   - **C** gene therapy
   - **D** tissue culture

Score /5

Short-answer questions

1. The diagram shows how Dolly the sheep was cloned.

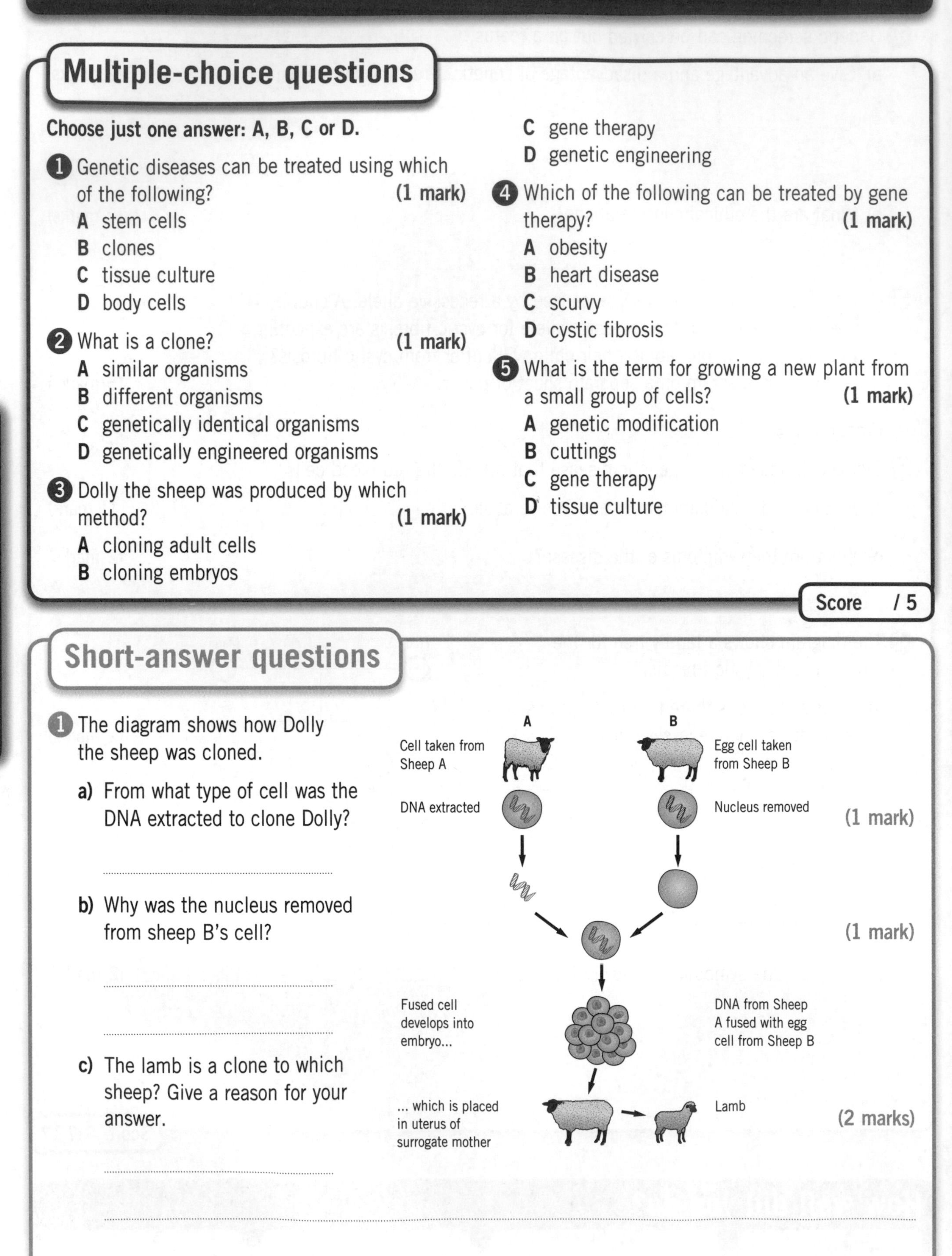

   a) From what type of cell was the DNA extracted to clone Dolly? **(1 mark)**

   ....

   b) Why was the nucleus removed from sheep B's cell? **(1 mark)**

   ....

   ....

   c) The lamb is a clone to which sheep? Give a reason for your answer. **(2 marks)**

   ....

   ....

Score /4

GCSE-style questions

Answer all parts of all questions. Continue on a separate sheet of paper if necessary.

1. Describe how plants can be cloned from cuttings and from small groups of cells. **(3 marks)**

2. The following sentences describe the process of genetic engineering, but they are in the wrong order. Put them in the correct order in the boxes provided. **(5 marks)**

   A The gene is cut using an enzyme.

   B The plasmid is cut using an enzyme and the gene is inserted.

   C The gene is identified from the strand of DNA.

   D The bacteria multiply to form clones.

   E The plasmid is put back in the bacterium.

3. **a)** What are stem cells? **(1 mark)**

   b) How could stem cells be used by doctors? **(1 mark)**

4. Why are identical twins clones? **(2 marks)**

5. **a)** Give two examples of why crops may be genetically modified. **(2 marks)**

   b) GM crops are very controversial. Give two arguments for GM crops and two against GM crops. (Answer on a separate sheet of paper.) **(4 marks)**

Score / 18

How well did you do?

0–6	Try again	7–12	Getting there	13–19	Good work	20–27	Excellent!

For more information on this topic, see pages 38–39 of your Success Revision Guide.

Evolution and Natural Selection

Multiple-choice questions

Choose just one answer: A, B, C or D.

1. What evidence is there for evolution? **(1 mark)**
   A photos
   B documents
   C fossils
   D plants

2. Whose theory of evolution is generally believed? **(1 mark)**
   A Darwin
   B Lamarck
   C Mendel
   D Newton

3. Natural selection can be considered as which of the following? **(1 mark)**
   A survival of the fittest
   B variation
   C selective breeding
   D artificial selection

4. Which of the following is an example of natural selection? **(1 mark)**
   A GM crops
   B Dolly the sheep
   C antibiotic-resistant bacteria
   D genetically modifying bacteria to produce insulin

5. Following the industrial revolution, why were more dark peppered moths found? **(1 mark)**
   A flew away from predators faster
   B protected species
   C better camouflaged from predators
   D poisonous to predators

Score /5

Short-answer questions

1. What is evolution? (2 marks)

   ....

   ....

2. Complete the following. (5 marks)

   Darwin called his theory .... It was based on these four observations:

   a) Organisms produce .... numbers of offspring.

   b) Population numbers .... over long time periods.

   c) Organisms of the same species are all slightly different: they show ....

   d) These beneficial characteristics can be .... from their parents.

Score /7

GCSE-style questions

Answer all parts of all questions. Continue on a separate sheet of paper if necessary.

1. Lamarck's theory states that giraffes' long necks developed due to them stretching for food. What evidence is there against his theory? **(2 marks)**

2. Why do some organisms become extinct? **(2 marks)**

3. Peppered moths evolved so that in industrial areas a dark variety was more prominent.

   a) Why are there more dark peppered moths in industrial areas? **(2 marks)**

   b) What caused the change in the peppered moths' colour? **(1 mark)**

   c) Use Darwin's theory to explain why the population of dark peppered moths increased. **(4 marks)**

Score / 11

How well did you do?

0–5	Try again	6–11	Getting there	12–17	Good work	18–23	Excellent!

For more information on this topic, see pages 40–41 of your Success Revision Guide.

Classifying Organisms

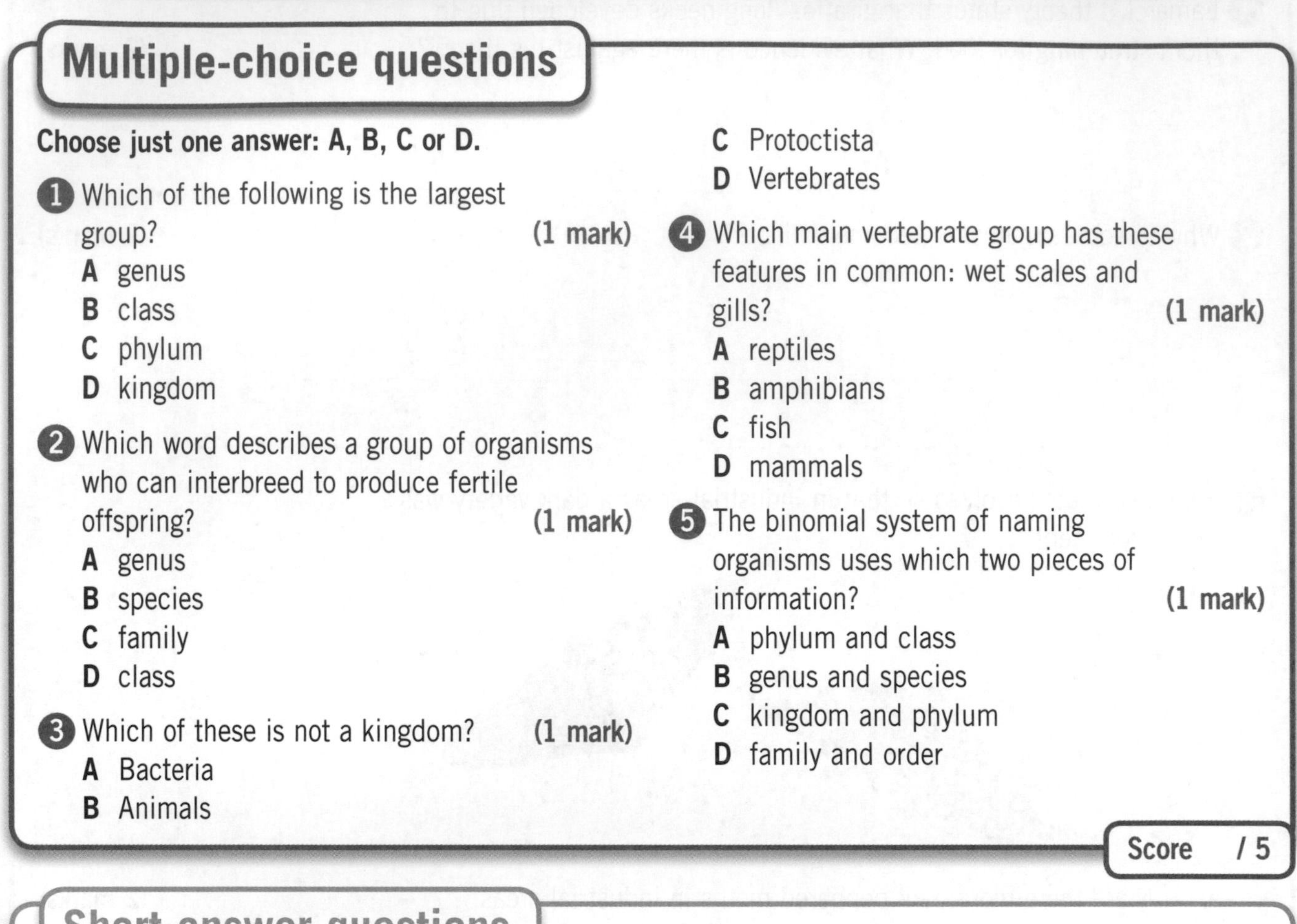

Multiple-choice questions

Choose just one answer: A, B, C or D.

1. Which of the following is the largest group? **(1 mark)**
   - **A** genus
   - **B** class
   - **C** phylum
   - **D** kingdom

2. Which word describes a group of organisms who can interbreed to produce fertile offspring? **(1 mark)**
   - **A** genus
   - **B** species
   - **C** family
   - **D** class

3. Which of these is not a kingdom? **(1 mark)**
   - **A** Bacteria
   - **B** Animals
   - **C** Protoctista
   - **D** Vertebrates

4. Which main vertebrate group has these features in common: wet scales and gills? **(1 mark)**
   - **A** reptiles
   - **B** amphibians
   - **C** fish
   - **D** mammals

5. The binomial system of naming organisms uses which two pieces of information? **(1 mark)**
   - **A** phylum and class
   - **B** genus and species
   - **C** kingdom and phylum
   - **D** family and order

Score / 5

Short-answer questions

1. Name a difference between the cells found in a plant and a fungus. (1 mark)

   ....

2. Bacteria were originally grouped with the kingdom of Protoctista. Look at the arguments that some scientists made when discussing if they should be considered the same kingdom or different kingdoms. For each argument, tick the box to indicate what the evidence supports. (4 marks)

Argument	Different?	Alike?
a) Bacteria have no true nucleus; protoctists have a nucleus		
b) Bacteria produce asexually; many protoctists can reproduce sexually		
c) Some bacteria and some protoctists can photosynthesise		
d) Bacteria and protoctists are unicellular or simple organisms		

3. Circle the information that is least likely to be useful in classifying an animal.

   habitat **anatomy** **fossil record** **DNA and biochemical analysis** (1 mark)

Score / 6

GCSE-style questions

Answer all parts of all questions. Continue on a separate sheet of paper if necessary.

1. Birds and reptiles are thought to have more common ancestry than the other groups of organisms.

   a) What piece of biotechnological information has confirmed this for scientists? **(1 mark)**

   ....

   b) Before this was available, give one example of other evidence that would have been used to help classify organisms. **(1 mark)**

   ....

   c) Read the paragraph about the vertebrate groups.

   > Fish, reptiles and amphibians are cold-blooded whereas birds and mammals are warm-blooded. Mammals, birds, reptiles and adult amphibians have lungs. Some amphibians such as baby frogs (tadpoles) have gills rather than lungs. Fish have gills. The skin of mammals is covered with hair or fur, whereas that of both reptiles and fish has scales. Birds too, have some areas of scales on their skin but, commonly, they have feathers and wings.

   From the text, identify one similarity which could indicate a common ancestry in birds and reptiles. **(1 mark)**

   ....

2. Tick the box that indicates the **best** observation regarding the usefulness of knowing an animal's habitat when classifying it. **(1 mark)**

   - [ ] Habitat of an animal is crucial in identifying organisms of the same species.
   - [ ] Habitat can be different for adults and young of the same species.
   - [ ] Living in the same habitat is not an indication that organisms are related.
   - [ ] Animals of the same species can exploit more than one habitat.

3. A lion and a tiger can mate to produce a liger. Are ligers likely to be fertile? Explain your answer. **(2 marks)**

   ....

   ....

4. Which vertebrate class do bats belong to? Justify your answer. **(2 marks)**

   ....

Score /8

How well did you do?

0–5	Try again	6–10	Getting there	11–14	Good work	15–19	Excellent!

For more information on this topic, see pages 44–45 of your Success Revision Guide.

Competition and Adaptation

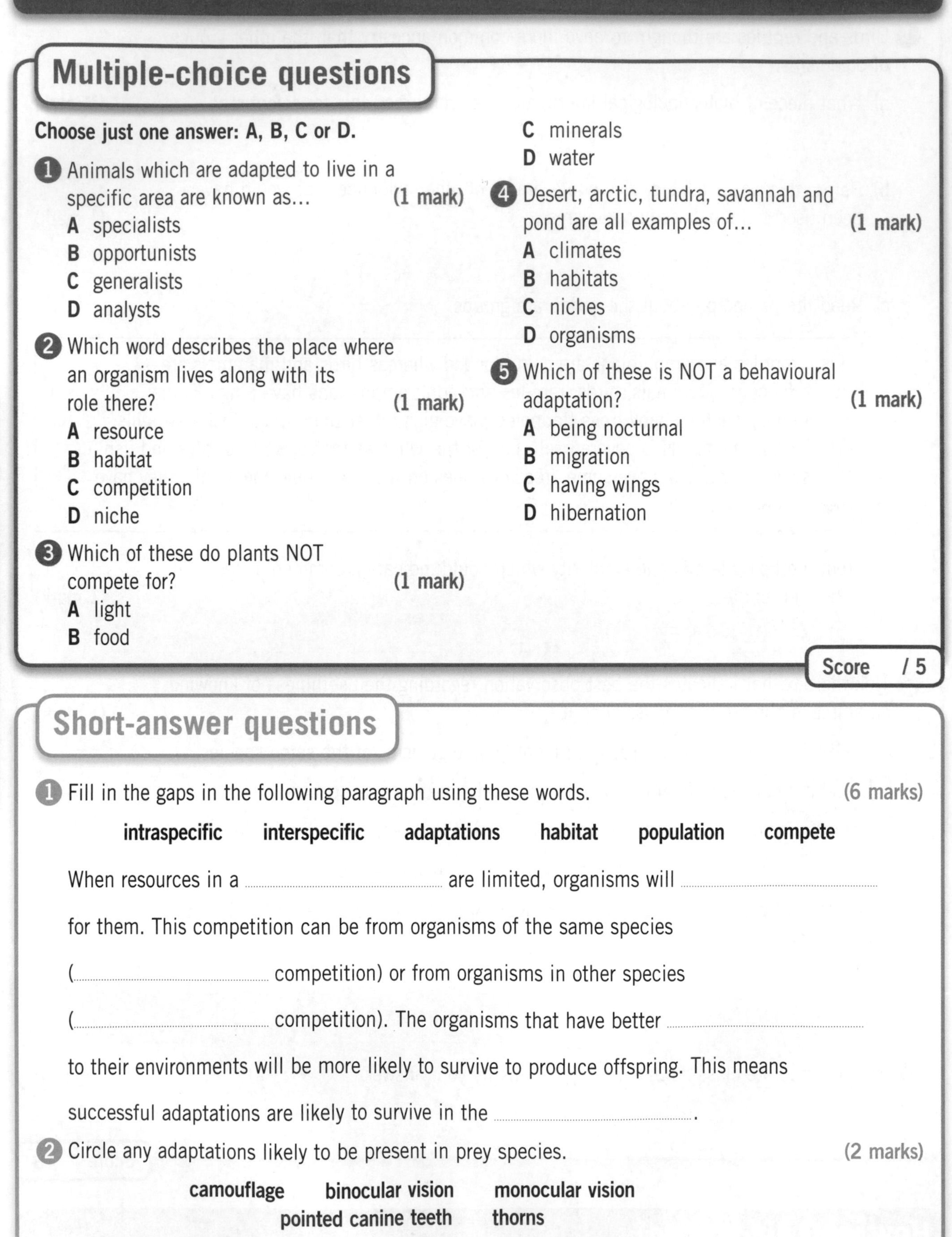

Multiple-choice questions

Choose just one answer: A, B, C or D.

1. Animals which are adapted to live in a specific area are known as... **(1 mark)**
   A specialists
   B opportunists
   C generalists
   D analysts

2. Which word describes the place where an organism lives along with its role there? **(1 mark)**
   A resource
   B habitat
   C competition
   D niche

3. Which of these do plants NOT compete for? **(1 mark)**
   A light
   B food
   C minerals
   D water

4. Desert, arctic, tundra, savannah and pond are all examples of... **(1 mark)**
   A climates
   B habitats
   C niches
   D organisms

5. Which of these is NOT a behavioural adaptation? **(1 mark)**
   A being nocturnal
   B migration
   C having wings
   D hibernation

Score /5

Short-answer questions

1. Fill in the gaps in the following paragraph using these words. **(6 marks)**

   intraspecific interspecific adaptations habitat population compete

   When resources in a .... are limited, organisms will .... for them. This competition can be from organisms of the same species (.................... competition) or from organisms in other species (.................... competition). The organisms that have better .... to their environments will be more likely to survive to produce offspring. This means successful adaptations are likely to survive in the ....

2. Circle any adaptations likely to be present in prey species. **(2 marks)**

   camouflage binocular vision monocular vision pointed canine teeth thorns

Score /8

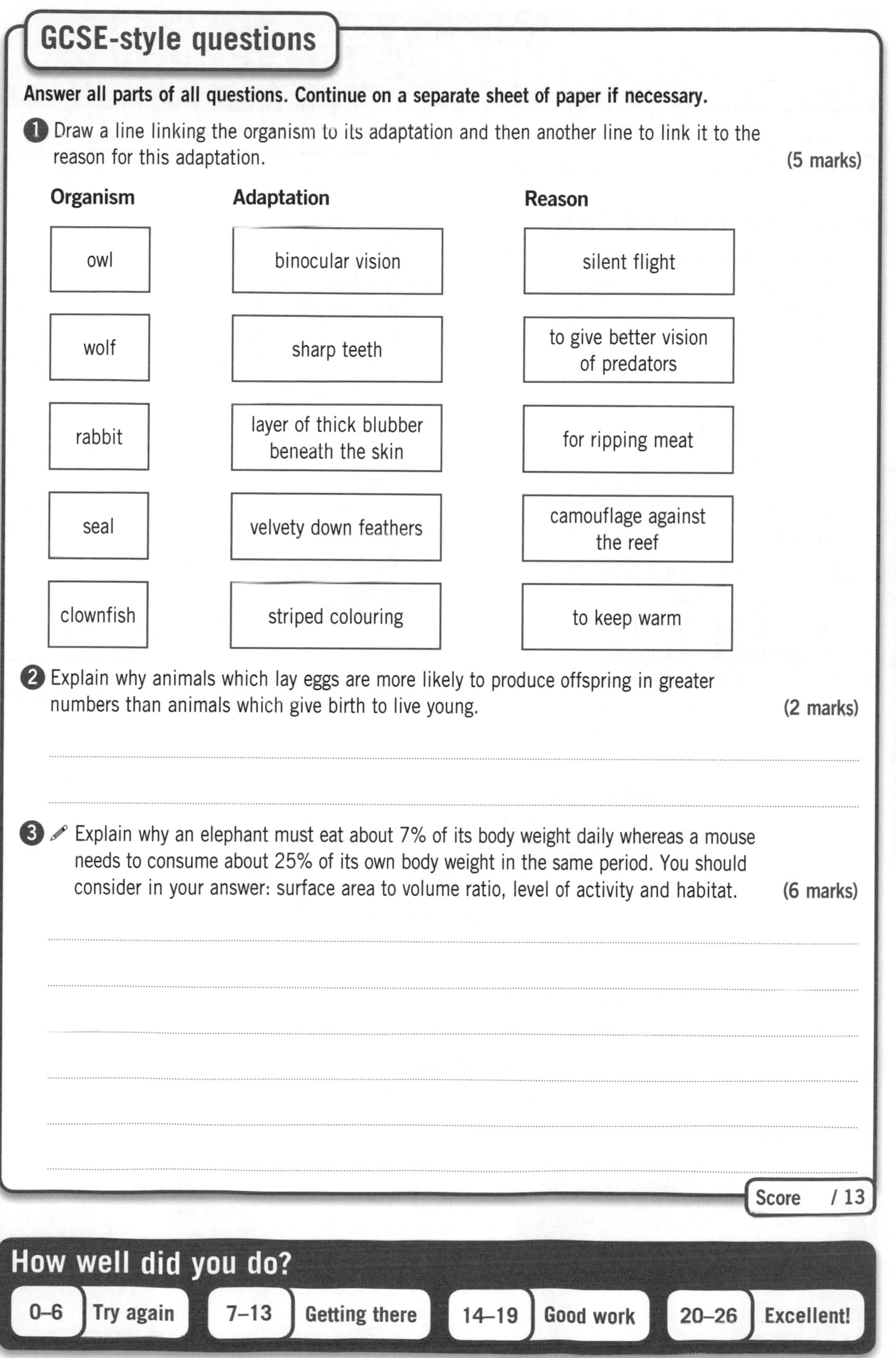

GCSE-style questions

Answer all parts of all questions. Continue on a separate sheet of paper if necessary.

1. Draw a line linking the organism to its adaptation and then another line to link it to the reason for this adaptation. **(5 marks)**

Organism	**Adaptation**	**Reason**
owl	binocular vision	silent flight
wolf	sharp teeth	to give better vision of predators
rabbit	layer of thick blubber beneath the skin	for ripping meat
seal	velvety down feathers	camouflage against the reef
clownfish	striped colouring	to keep warm

2. Explain why animals which lay eggs are more likely to produce offspring in greater numbers than animals which give birth to live young. **(2 marks)**

3. Explain why an elephant must eat about 7% of its body weight daily whereas a mouse needs to consume about 25% of its own body weight in the same period. You should consider in your answer: surface area to volume ratio, level of activity and habitat. **(6 marks)**

Score / 13

How well did you do?

0–6 Try again | 7–13 Getting there | 14–19 Good work | 20–26 Excellent!

For more information on this topic, see pages 46–47 of your Success Revision Guide.

Living Together

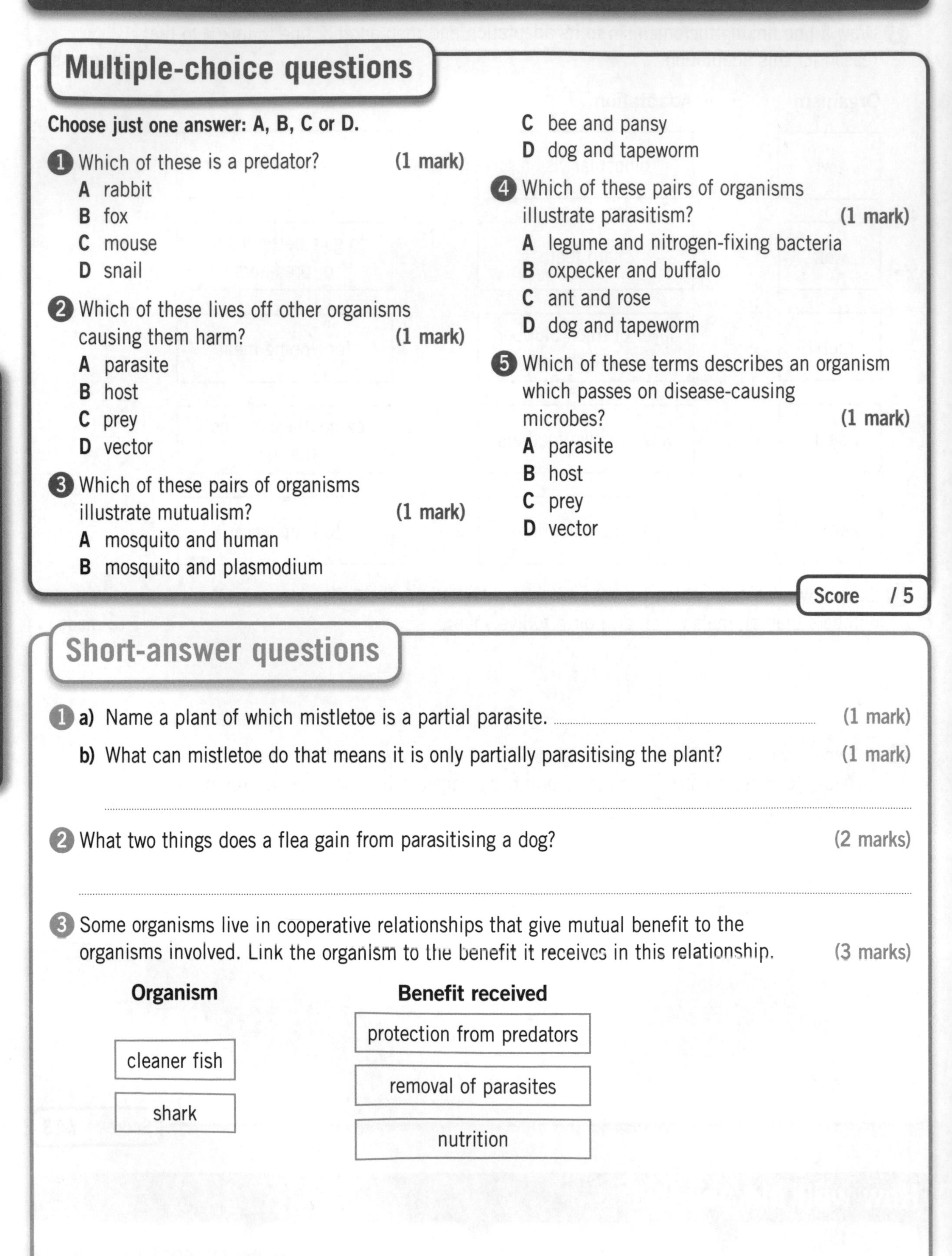

Multiple-choice questions

Choose just one answer: A, B, C or D.

1. Which of these is a predator? **(1 mark)**
   - **A** rabbit
   - **B** fox
   - **C** mouse
   - **D** snail

2. Which of these lives off other organisms causing them harm? **(1 mark)**
   - **A** parasite
   - **B** host
   - **C** prey
   - **D** vector

3. Which of these pairs of organisms illustrate mutualism? **(1 mark)**
   - **A** mosquito and human
   - **B** mosquito and plasmodium
   - **C** bee and pansy
   - **D** dog and tapeworm

4. Which of these pairs of organisms illustrate parasitism? **(1 mark)**
   - **A** legume and nitrogen-fixing bacteria
   - **B** oxpecker and buffalo
   - **C** ant and rose
   - **D** dog and tapeworm

5. Which of these terms describes an organism which passes on disease-causing microbes? **(1 mark)**
   - **A** parasite
   - **B** host
   - **C** prey
   - **D** vector

Score /5

Short-answer questions

1. **a)** Name a plant of which mistletoe is a partial parasite. .... (1 mark)

   b) What can mistletoe do that means it is only partially parasitising the plant? (1 mark)

   ....

2. What two things does a flea gain from parasitising a dog? (2 marks)

   ....

3. Some organisms live in cooperative relationships that give mutual benefit to the organisms involved. Link the organism to the benefit it receives in this relationship. (3 marks)

Organism	**Benefit received**
cleaner fish	protection from predators
shark	removal of parasites
	nutrition

Score /7

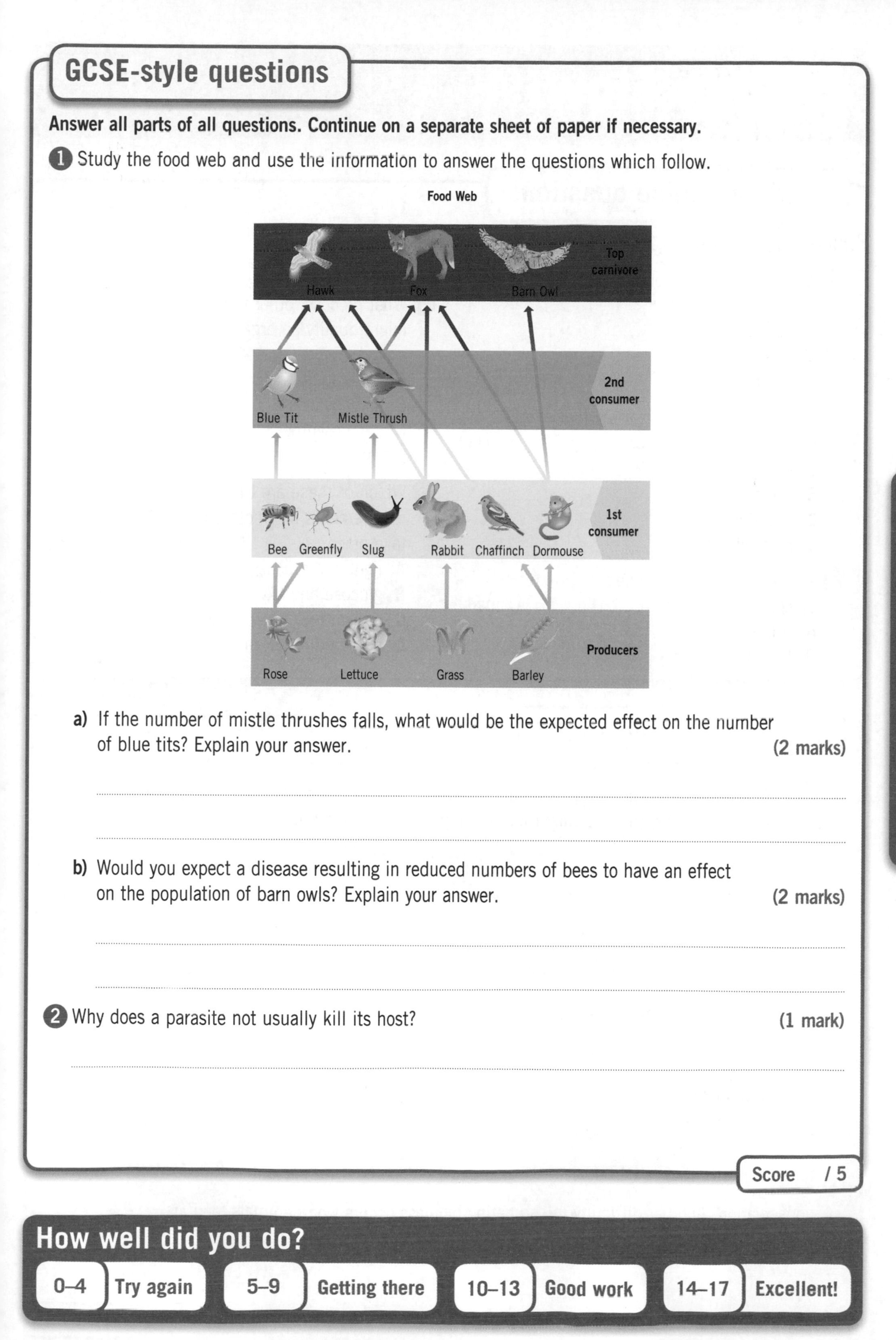

GCSE-style questions

Answer all parts of all questions. Continue on a separate sheet of paper if necessary.

1 Study the food web and use the information to answer the questions which follow.

a) If the number of mistle thrushes falls, what would be the expected effect on the number of blue tits? Explain your answer. **(2 marks)**

....

....

b) Would you expect a disease resulting in reduced numbers of bees to have an effect on the population of barn owls? Explain your answer. **(2 marks)**

....

....

2 Why does a parasite not usually kill its host? **(1 mark)**

....

Score / 5

How well did you do?

0–4	Try again	5–9	Getting there	10–13	Good work	14–17	Excellent!

For more information on this topic, see pages 48–49 of your Success Revision Guide.

Energy Flow

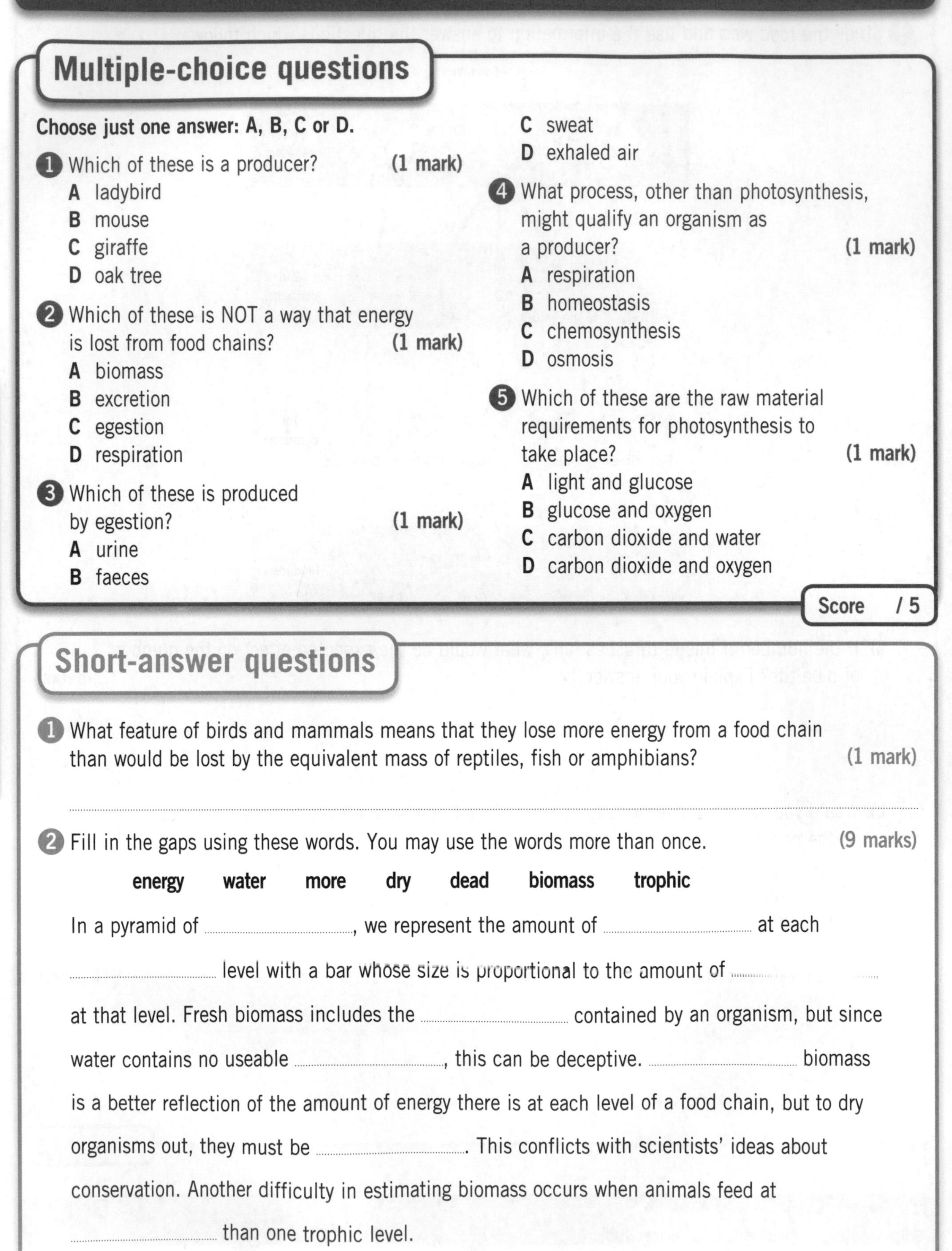

Multiple-choice questions

Choose just one answer: A, B, C or D.

1. Which of these is a producer? **(1 mark)**
   - **A** ladybird
   - **B** mouse
   - **C** giraffe
   - **D** oak tree

2. Which of these is NOT a way that energy is lost from food chains? **(1 mark)**
   - **A** biomass
   - **B** excretion
   - **C** egestion
   - **D** respiration

3. Which of these is produced by egestion? **(1 mark)**
   - **A** urine
   - **B** faeces
   - **C** sweat
   - **D** exhaled air

4. What process, other than photosynthesis, might qualify an organism as a producer? **(1 mark)**
   - **A** respiration
   - **B** homeostasis
   - **C** chemosynthesis
   - **D** osmosis

5. Which of these are the raw material requirements for photosynthesis to take place? **(1 mark)**
   - **A** light and glucose
   - **B** glucose and oxygen
   - **C** carbon dioxide and water
   - **D** carbon dioxide and oxygen

Score /5

Short-answer questions

1. What feature of birds and mammals means that they lose more energy from a food chain than would be lost by the equivalent mass of reptiles, fish or amphibians? (1 mark)

   ....

2. Fill in the gaps using these words. You may use the words more than once. (9 marks)

   energy   water   more   dry   dead   biomass   trophic

   In a pyramid of ...., we represent the amount of .... at each .... level with a bar whose size is proportional to the amount of .... at that level. Fresh biomass includes the .... contained by an organism, but since water contains no useable ...., this can be deceptive. .... biomass is a better reflection of the amount of energy there is at each level of a food chain, but to dry organisms out, they must be .... This conflicts with scientists' ideas about conservation. Another difficulty in estimating biomass occurs when animals feed at .... than one trophic level.

Score /10

GCSE-style questions

Answer all parts of all questions. Continue on a separate sheet of paper if necessary.

1. Some farmers choose to restrict the movement of their animals to maximise meat production. Explain how this works. **(2 marks)**

2. Read the information below. Use this and your knowledge of food chains and pyramids of biomass to answer the questions that follow.

In the 1960s, a pesticide called DDT was widely used to increase crop yield. It worked by poisoning the insects that ate the crops. Once inside the body of an animal, DDT takes a very long time to break down.

Pesticide which fell onto the soil was washed into water courses in rainwater. Environmentalists saw a correlation in the use of DDT and the decline in the number of grebes and fish in affected areas. They lobbied for a change in the law and DDT was withdrawn from use.

a) Explain why the poison killed more grebes than it did fish. **(4 marks)**

b) What is meant by the term 'correlation'? **(1 mark)**

c) What evidence could the environmentalists have provided to support their view? **(3 marks)**

3. Using the food web on page 43, explain why the number of barn owls is more likely to be affected than the number of foxes, if the number of dormice falls. **(2 marks)**

Score / 12

How well did you do?

0–6	Try again	7–12	Getting there	13–19	Good work	20–27	Excellent!

For more information on this topic, see pages 50–51 of your Success Revision Guide.

Recycling

Multiple-choice questions

Choose just one answer: A, B, C or D.

1. Which of these is a decomposer? **(1 mark)**
   A fungi
   B grass
   C oak
   D corn

2. Which of these conditions will help prevent decay by (most) bacteria? **(1 mark)**
   A well oxygenated
   B acidic
   C damp
   D warm

3. What name is given to decaying plant matter that gardeners may add to fertilise their soil? **(1 mark)**
   A compost
   B mess
   C decomposers
   D fungi

4. Decomposers are useful to plants because they recycle what back into the soil? **(1 mark)**
   A water
   B light
   C minerals
   D carbon dioxide

5. Which chemical do nitrifying bacteria convert to nitrogen? **(1 mark)**
   A ammonia
   B nitrogen
   C oxygen
   D carbon dioxide

Score /5

Short-answer questions

1. Name three processes that release CO_2 into the air. .... (3 marks)

....

2. The arrows in the picture show carbon transfer in the carbon cycle. Match the letters A–F with the processes. (6 marks)

	death
	feeding
	combustion
	decomposition
	respiration
	photosynthesis

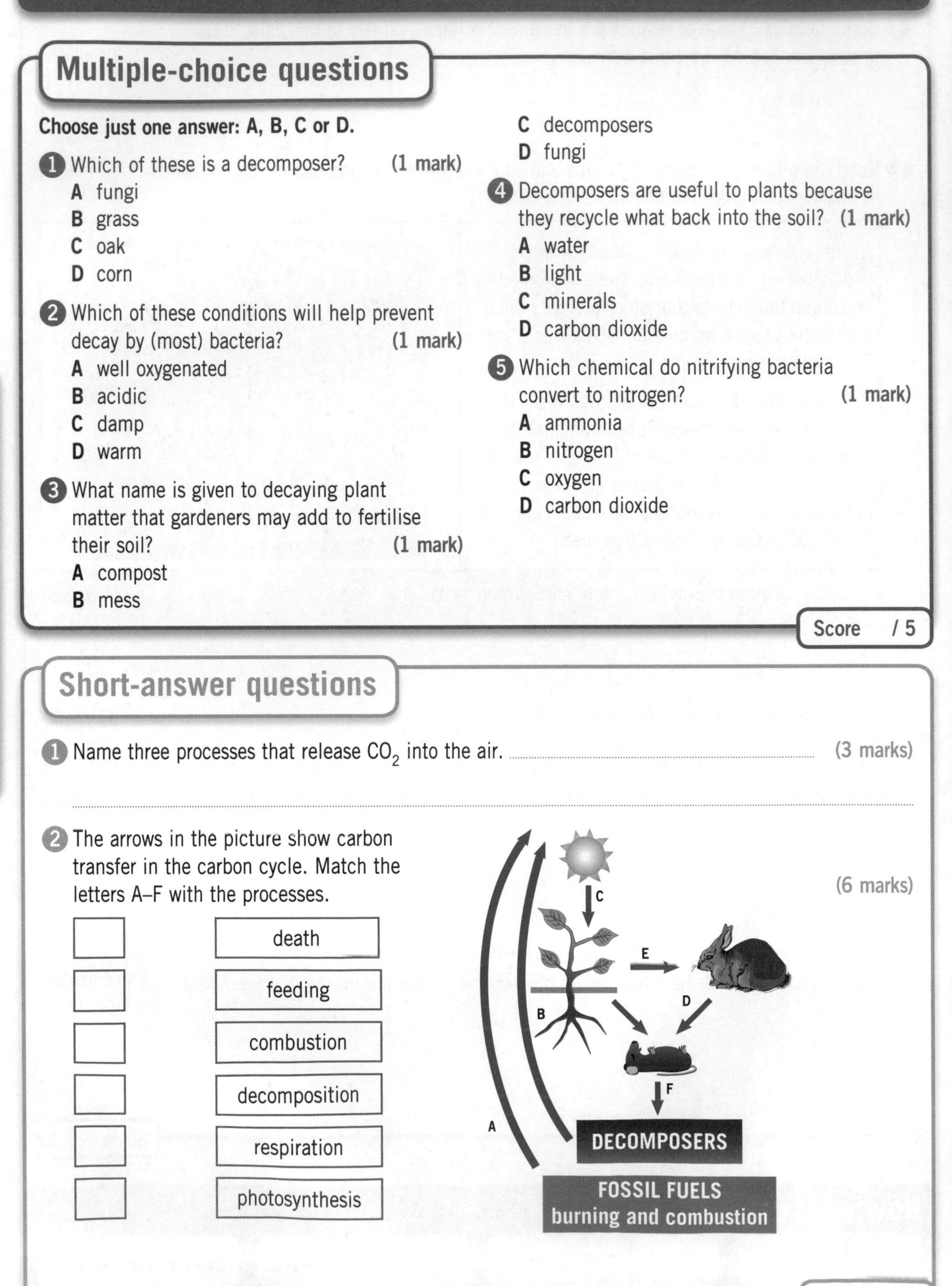

Score /9

GCSE-style questions

Answer all parts of all questions. Continue on a separate sheet of paper if necessary.

1 These questions are about the nitrogen cycle.

a) What organisms are responsible for converting nitrogen in the soil into nitrates that are useable by the plant? .... **(1 mark)**

b) Why is nitrogen unusable whereas nitrates are useable? **(2 marks)**

....

c) What role does lightning play in the nitrogen cycle? **(1 mark)**

....

2 The graph shows what happens to the rate at which bacteria work at different temperatures.

Rate of decay

P

Q

0

45°C

Temperature

a) Describe the relationship between temperature and rate of decay. **(3 marks)**

....

....

....

b) Explain the relationship you have described. **(2 marks)**

....

....

c) Related to this correlation, explain why it is necessary to turn a compost heap over in order to keep the decay continuous. **(3 marks)**

....

....

....

3 What would happen if there were no decomposers in an ecosystem? **(1 mark)**

....

....

Score / 13

How well did you do?

0–6	Try again	7–12	Getting there	13–19	Good work	20–27	Excellent!

For more information on this topic, see pages 52–53 of your Success Revision Guide.

Populations and Pollution

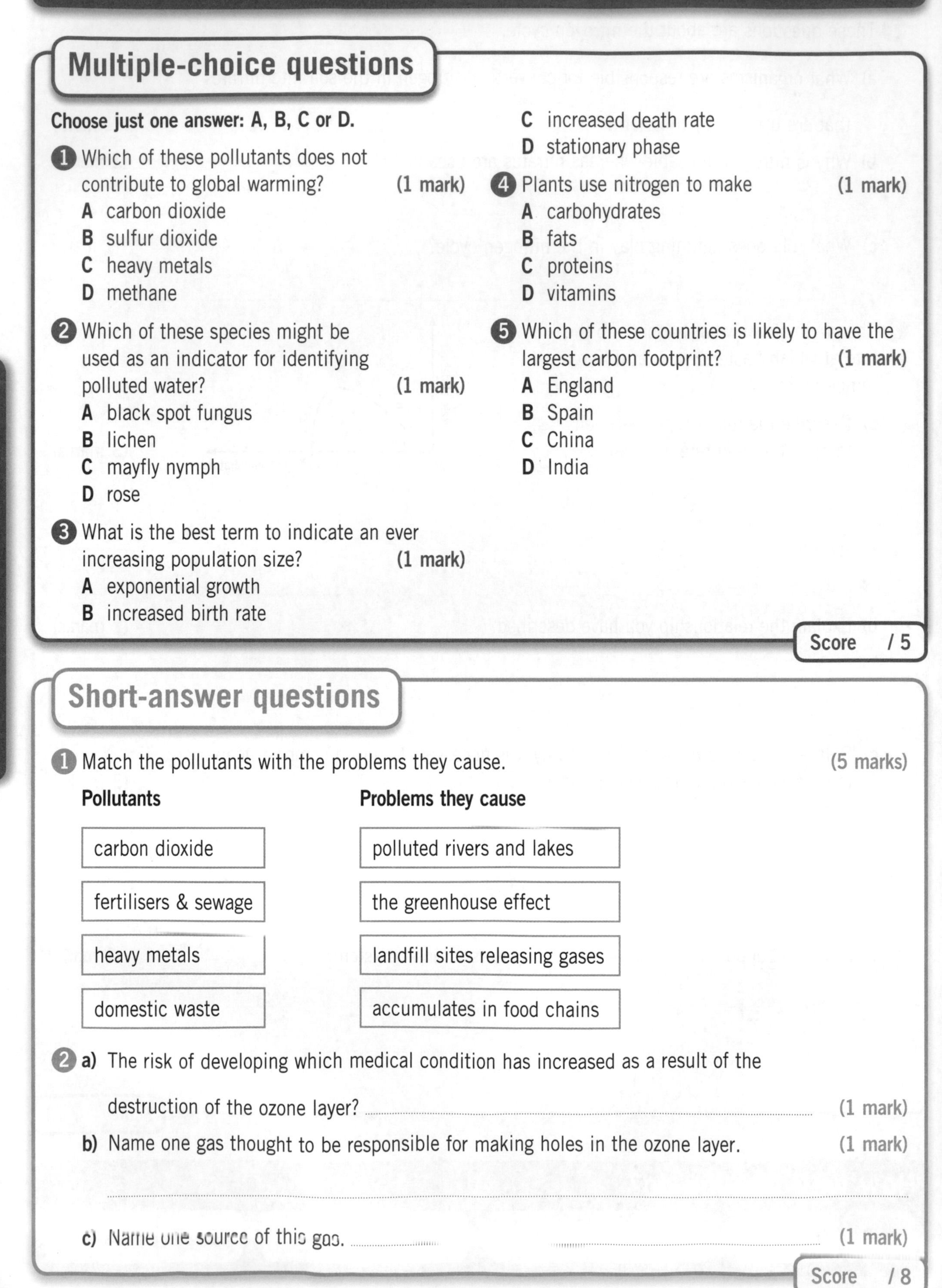

Multiple-choice questions

Choose just one answer: A, B, C or D.

1. Which of these pollutants does not contribute to global warming? **(1 mark)**
   - **A** carbon dioxide
   - **B** sulfur dioxide
   - **C** heavy metals
   - **D** methane

2. Which of these species might be used as an indicator for identifying polluted water? **(1 mark)**
   - **A** black spot fungus
   - **B** lichen
   - **C** mayfly nymph
   - **D** rose

3. What is the best term to indicate an ever increasing population size? **(1 mark)**
   - **A** exponential growth
   - **B** increased birth rate
   - **C** increased death rate
   - **D** stationary phase

4. Plants use nitrogen to make **(1 mark)**
   - **A** carbohydrates
   - **B** fats
   - **C** proteins
   - **D** vitamins

5. Which of these countries is likely to have the largest carbon footprint? **(1 mark)**
   - **A** England
   - **B** Spain
   - **C** China
   - **D** India

Score /5

Short-answer questions

1. Match the pollutants with the problems they cause. (5 marks)

Pollutants	Problems they cause
carbon dioxide	polluted rivers and lakes
fertilisers & sewage	the greenhouse effect
heavy metals	landfill sites releasing gases
domestic waste	accumulates in food chains

2. **a)** The risk of developing which medical condition has increased as a result of the destruction of the ozone layer? .... (1 mark)

   b) Name one gas thought to be responsible for making holes in the ozone layer. (1 mark)

   ....

   c) Name one source of this gas. .... (1 mark)

Score /8

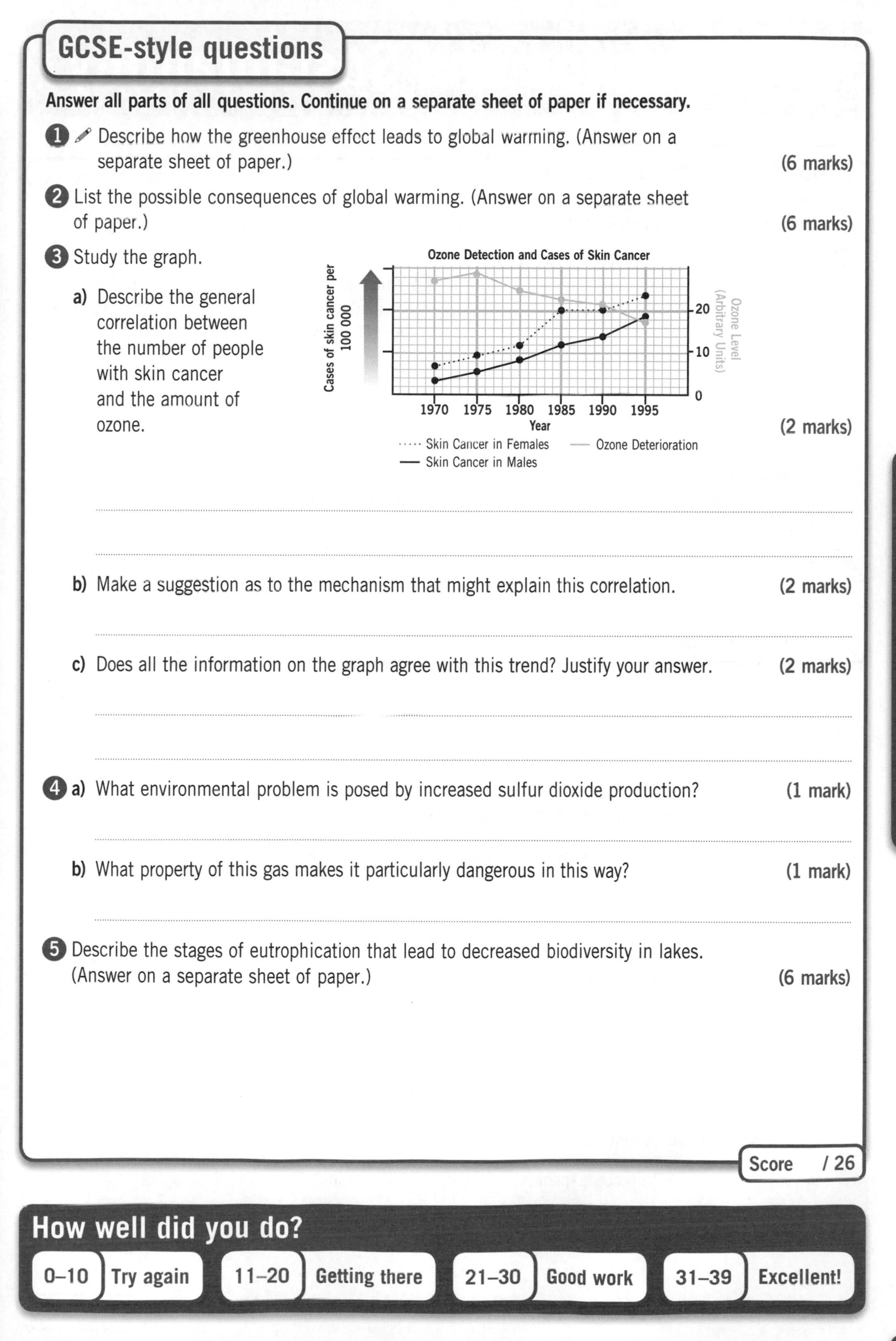

GCSE-style questions

Answer all parts of all questions. Continue on a separate sheet of paper if necessary.

1. Describe how the greenhouse effect leads to global warming. (Answer on a separate sheet of paper.) **(6 marks)**

2. List the possible consequences of global warming. (Answer on a separate sheet of paper.) **(6 marks)**

3. Study the graph.

   a) Describe the general correlation between the number of people with skin cancer and the amount of ozone. **(2 marks)**

   ....

   ....

   b) Make a suggestion as to the mechanism that might explain this correlation. **(2 marks)**

   ....

   c) Does all the information on the graph agree with this trend? Justify your answer. **(2 marks)**

   ....

   ....

4. **a)** What environmental problem is posed by increased sulfur dioxide production? **(1 mark)**

   ....

   b) What property of this gas makes it particularly dangerous in this way? **(1 mark)**

   ....

5. Describe the stages of eutrophication that lead to decreased biodiversity in lakes. (Answer on a separate sheet of paper.) **(6 marks)**

Score / 26

How well did you do?

0–10 Try again | 11–20 Getting there | 21–30 Good work | 31–39 Excellent!

For more information on this topic, see pages 54–55 of your Success Revision Guide.

Conservation and Sustainability

Multiple-choice questions

Choose just one answer: A, B, C, or D.

1. What word describes 'a place to live'? **(1 mark)**
   A ecosystem
   B habitat
   C climate
   D competition
2. What word means the 'variety of different living organisms in a habitat'? **(1 mark)**
   A biodiversity
   B population
   C community
   D ecosystem
3. Which of these might contribute to an organism becoming endangered? **(1 mark)**
   A competition
   B climate change
   C habitat destruction
   D all of these
4. Which of these hopes to maintain species diversity? **(1 mark)**
   A deforestation
   B soil erosion
   C intensive farming
   D organic farming
5. Trying to maintain habitats and the species that live in them is the aim of **(1 mark)**
   A conservation
   B deforestation
   C land clearance
   D high-level fishing

Score /5

Short-answer questions

1. Use these words to fill in the gaps in the following paragraph. (7 marks)

compete **laws** **endangered** **education** **pollutants** **extinct** **hunted**

Human activity can lead to species becoming .... or ....

For example, whales are .... by humans for food and oil. This is an example where humans are intentionally responsible for reducing whale numbers but humans can impact the whale population in other ways. Humans using boats might accidentally hit the whales in shipping channels. These boats also produce ...., which can damage the whales' habitat. Disturbances to the sea bed might reduce the amount of food available to the whales. Additionally, humans .... for the whales' food by fishing in their habitat. In order to reduce our impact on the environment we have passed .... and have invested in .... to protect species and maintain biodiversity.

Score /7

GCSE-style questions

Answer all parts of all questions. Continue on a separate sheet of paper if necessary.

1

Distance from city centre	Number of different lichen species	Sulfur dioxide levels (ppm)
0	0	210
2	1	144
4	4	91
6	8	47
8	12	10

a) Plot a graph of the data about sulfur levels at different distances from the city centre. Add a trend line. **(3 marks)**

b) You are going to add more data to the same graph using the second, right-hand scale. Use the table to add plots to represent the number of different lichen species at different distances from the city centre. Add a trend line for this. **(3 marks)**

Sulfur dioxide levels (ppm)

Number of different lichen species

Distance from the city centre

c) Describe the trend shown by the data between the sulfur dioxide level and the number of different species of lichens seen. **(1 mark)**

....

....

d) If pollution is sustained over a long period of time, what might be the expected outcome for a population with:

i) large genetic variation in its individuals? **(1 mark)**

....

ii) limited genetic variation in its individuals? **(1 mark)**

....

Score /9

How well did you do?

0–5 Try again | 6–10 Getting there | 11–15 Good work | 16–21 Excellent!

For more information on this topic, see pages 56–57 of your Success Revision Guide.

Cells and Organisation

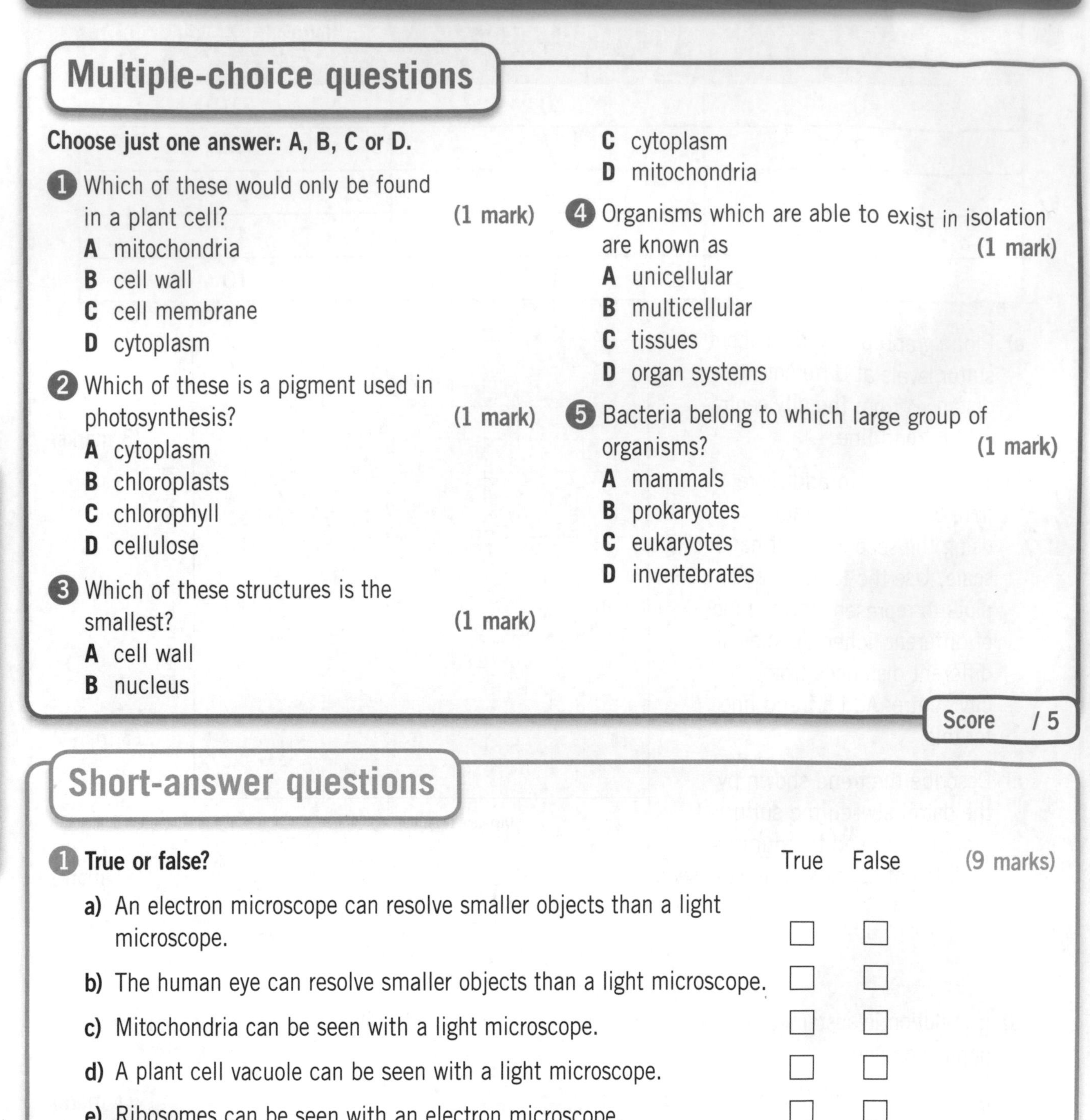

Multiple-choice questions

Choose just one answer: A, B, C or D.

1. Which of these would only be found in a plant cell? **(1 mark)**
   A mitochondria
   B cell wall
   C cell membrane
   D cytoplasm

2. Which of these is a pigment used in photosynthesis? **(1 mark)**
   A cytoplasm
   B chloroplasts
   C chlorophyll
   D cellulose

3. Which of these structures is the smallest? **(1 mark)**
   A cell wall
   B nucleus
   C cytoplasm
   D mitochondria

4. Organisms which are able to exist in isolation are known as **(1 mark)**
   A unicellular
   B multicellular
   C tissues
   D organ systems

5. Bacteria belong to which large group of organisms? **(1 mark)**
   A mammals
   B prokaryotes
   C eukaryotes
   D invertebrates

Score / 5

Short-answer questions

1. **True or false?** **(9 marks)**

	True	False
a) An electron microscope can resolve smaller objects than a light microscope.	☐	☐
b) The human eye can resolve smaller objects than a light microscope.	☐	☐
c) Mitochondria can be seen with a light microscope.	☐	☐
d) A plant cell vacuole can be seen with a light microscope.	☐	☐
e) Ribosomes can be seen with an electron microscope.	☐	☐
f) Electron microscopes can resolve structures at around 0.002 mm.	☐	☐
g) A tissue is bigger than a cell.	☐	☐
h) Xylem is an example of a tissue.	☐	☐
i) A heart is an example of an organ.	☐	☐

Score / 9

GCSE-style questions

Answer all parts of all questions. Continue on a separate sheet of paper if necessary.

1 a) Identify which is an animal cell and which is a plant cell on the diagram captions. **(2 marks)**

b) Complete the labels on the diagrams. **(12 marks)**

1 .... cell

A .... C ....

B .... D ....

E .... G ....

F .... H ....

2 .... cell

K ....

.... L

J ....

I ....

c) Describe the role of mitochondria, chloroplasts and cytoplasm in the cell. **(3 marks)**

....

....

....

2 Is a muscle a tissue or an organ? Explain your answer. **(2 marks)**

....

....

3 Name three difficulties of being multicellular which unicellular organisms would not experience. **(3 marks)**

....

....

....

Score / 22

How well did you do?

0–9	Try again	10–18	Getting there	19–27	Good work	28–36	Excellent!

For more information on this topic, see pages 60–61 of your Success Revision Guide.

DNA and Protein Synthesis

Multiple-choice questions

Choose just one answer: A, B, C or D.

1. How many bases are there in DNA? **(1 mark)**
   A 1
   B 2
   C 3
   D 4
2. Which base always pairs with 'A' in DNA? **(1 mark)**
   A A
   B T
   C C
   D G
3. Which scientist did not focus their work on DNA directly? **(1 mark)**
   A Rosalind Franklin
   B Erwin Chargaff
   C Charles Darwin
   D Frances Crick
4. How many strands does DNA have? **(1 mark)**
   A 1
   B 2
   C 3
   D 4
5. Which of these is the largest structure? **(1 mark)**
   A DNA
   B gene
   C base
   D triplet code

Score / 5

Short-answer questions

1. **a)** What is a sequence of three bases known as? .... (1 mark)
   b) How many amino acids would these three bases code for? .... (1 mark)
   c) What is the name of the process whereby amino acids are identified by the code and joined to make a polypeptide (protein)? .... (1 mark)
   d) What molecule 'reads' the code and conducts the process of adding amino acids? (1 mark)
   ....
   e) Where does translation happen? .... (1 mark)
   f) Name the difference between DNA and mRNA. .... (1 mark)
   ....
2. **True or false?**

	True	False	
a) The evidence for DNA structure came from microscope images.	☐	☐	(1 mark)
b) Maurice Wilkins and Rosalind Franklin used X-rays to investigate DNA structure.	☐	☐	(1 mark)
c) Watson and Crick were responsible for suggesting that DNA has a double helix structure.	☐	☐	(1 mark)

Score / 9

GCSE-style questions

Answer all parts of all questions. Continue on a separate sheet of paper if necessary.

1 In what order do these events happen to synthesise a protein? **(6 marks)**

A	Section of DNA containing a gene 'unzips'.
B	mRNA forms a single stranded DNA template (transcription).
C	Triplet codes of bases on the new strand are 'read' by tRNA and ribosomes (translation).
D	Ribosomes add amino acids according to the triplet code.
E	mRNA moves out of nucleus to the cytoplasm.
F	Amino acids form a (polypeptide) protein.
G	Free bases (on tRNA) pair with mRNA strand.

ANSWER:

A						

2 Why can mRNA leave the nucleus but DNA cannot? **(1 mark)**

....

3 Which of these is not made of protein? (Tick the box.) **(1 mark)**

	DNA
	enzymes
	haemoglobin
	cell wall (in a plant)

4 What is meant by the term helix? .... **(1 mark)**

5 What substances make up the two 'backbone' strands of the helix? **(2 marks)**

....

6 Arrange the following in order of magnitude from the largest to the smallest. **(4 marks)**

chromosome **base** **protein** **gene**

.... → .... → .... → ....

Score / 15

How well did you do?

0–7 Try again | 8–14 Getting there | 15–21 Good work | 22–29 Excellent!

For more information on this topic, see pages 62–63 of your Success Revision Guide.

Proteins and Enzymes

Multiple-choice questions

Choose just one answer: A, B, C or D.

1. The optimum temperature for enzymes that work in the human body is **(1 mark)**
   - **A** 27 °C
   - **B** 37 °C
   - **C** 47 °C
   - **D** 57 °C

2. Which temperature is likely to produce the lowest rate of reaction for respiration in humans? **(1 mark)**
   - **A** 27 °C
   - **B** 37 °C
   - **C** 47 °C
   - **D** 57 °C

3. What is the name of the part of the enzyme where the substrate fits into the molecule? **(1 mark)**
   - **A** active immunity
   - **B** active site
   - **C** active revision
   - **D** active area

4. Heat changes the shape of the enzyme molecule. When this happens we say that the enzyme has been **(1 mark)**
   - **A** killed
   - **B** denatured
   - **C** melted
   - **D** squashed

5. The molecule that the enzyme works on is known as the **(1 mark)**
   - **A** substitute
   - **B** reactant
   - **C** product
   - **D** substrate

Score /5

Short-answer questions

1. Name three processes that use enzymes. (3 marks)

2. Give four uses of proteins in the body and give an example for each. (4 marks)

3. 'Biological' washing powder contains enzymes used by our bodies to digest food.

   a) Why are enzymes included in the washing powder? (1 mark)

   b) If you are using 'biological' washing powder, the manufacturers recommend a maximum washing temperature of 40 °C. Explain why this is. (2 marks)

Score /10

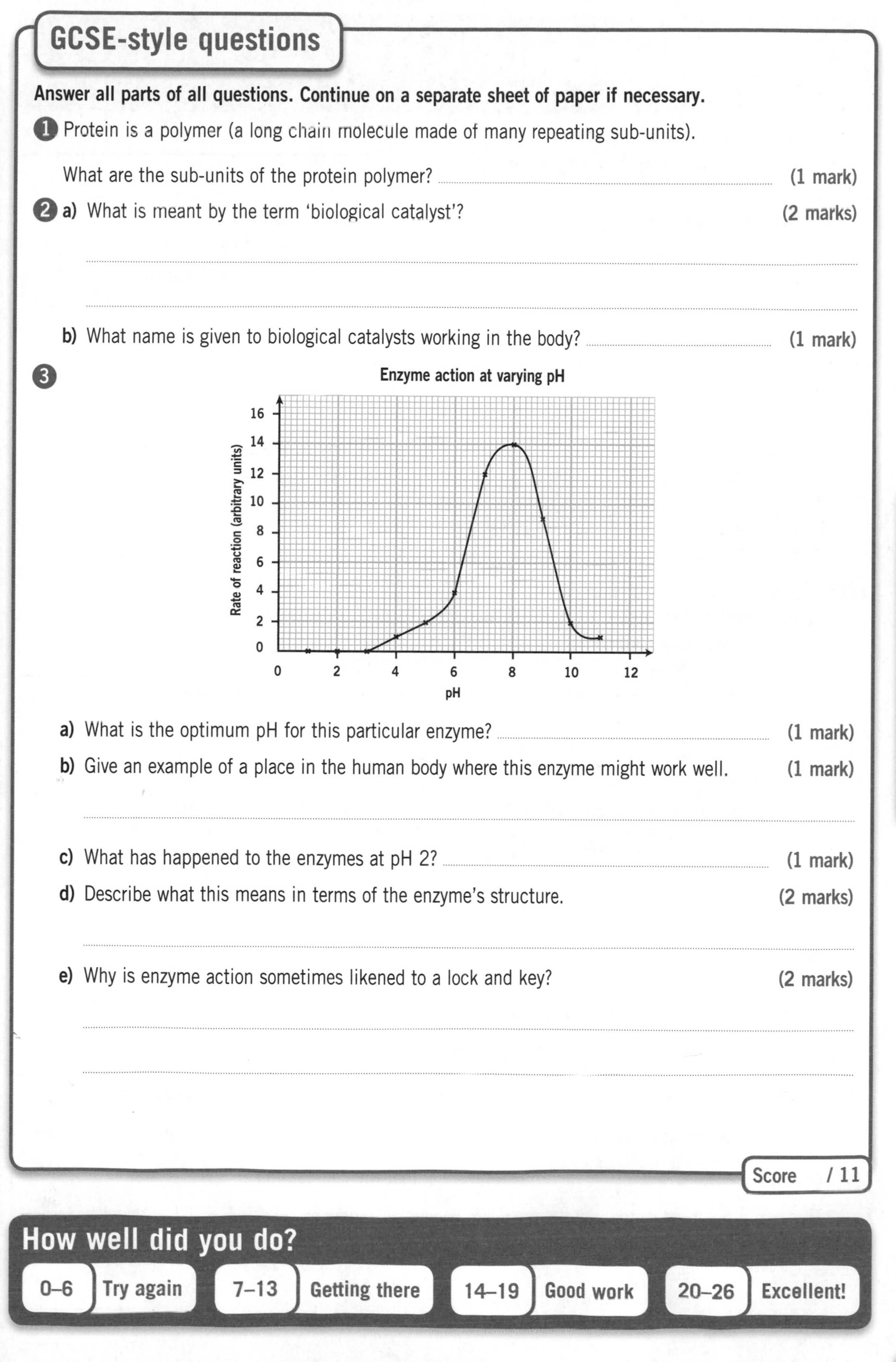

GCSE-style questions

Answer all parts of all questions. Continue on a separate sheet of paper if necessary.

1 Protein is a polymer (a long chain molecule made of many repeating sub-units).

What are the sub-units of the protein polymer? .... **(1 mark)**

2 a) What is meant by the term 'biological catalyst'? **(2 marks)**

....

....

b) What name is given to biological catalysts working in the body? .... **(1 mark)**

3

a) What is the optimum pH for this particular enzyme? .... **(1 mark)**

b) Give an example of a place in the human body where this enzyme might work well. **(1 mark)**

....

c) What has happened to the enzymes at pH 2? .... **(1 mark)**

d) Describe what this means in terms of the enzyme's structure. **(2 marks)**

....

e) Why is enzyme action sometimes likened to a lock and key? **(2 marks)**

....

....

Score / 11

How well did you do?

0–6 Try again | 7–13 Getting there | 14–19 Good work | 20–26 Excellent!

For more information on this topic, see pages 64–65 of your Success Revision Guide.

Cell Division

Multiple-choice questions

Choose just one answer: A, B, C or D.

1. Which of these is a type of cell division? **(1 mark)**
   A haploid
   B diploid
   C mitosis
   D mutation

2. Which of these is unlikely to cause a genetic mutation? **(1 mark)**
   A UV in sunlight
   B X-rays
   C ultrasound
   D chemical mutagens

3. Which of these does not result in variation in the offspring? **(1 mark)**
   A random pairing of parents
   B mutations in the genes
   C asexual reproduction
   D formation of gametes

4. How many cells are produced by mitotic cell division? **(1 mark)**
   A 1
   B 2
   C 3
   D 4

5. How many cells are produced by meiotic cell division? **(1 mark)**
   A 1
   B 2
   C 3
   D 4

Score / 5

Short-answer questions

1. **a)** What word describes a cell that has only one copy of each of its chromosomes? (1 mark)

   ....

   b) What name is given to these cells? .... (1 mark)

   c) What name is given to the process whereby two of these cells fuse? (1 mark)

   ....

   d) What type of reproduction involves this process? .... (1 mark)

2. Fill in the gaps in the paragraph below using these words. (7 marks)

   gene **DNA** **variation** **changes** **proteins** **insignificant** **mutations**

   Mutations can bring about changes to the .... of an organism. Sometimes this results in a new .... being formed, meaning that the cell can make new .... This might be an advantage for the organism because it produces .... Usually though, ... make such small .... to the DNA that they are .... in changing the gene.

Score / 11

Cells and Organisation

GCSE-style questions

Answer all parts of all questions. Continue on a separate sheet of paper if necessary.

1 **a)** What is the diploid number in human body cells? **(1 mark)**

....

b) What number of chromosomes would you find in a sperm cell? **(1 mark)**

....

c) The cells of which tissue type make sperm cells? **(1 mark)**

....

d) Which type of cell division is responsible for making the sperm? **(1 mark)**

....

e) Give one advantage of sexual reproduction over asexual reproduction. **(1 mark)**

....

2 What is the sequence of these events in meiosis? **(5 marks)**

A	Paired chromosomes separate.
B	Two cells are formed.
C	To form a total of four, new, haploid cells.
D	DNA becomes organised into chromosome pairs.
E	Each double stranded chromosome now separates.
F	DNA is copied.

F					

3 Mutation plays a great part in the evolution of strains of bacteria.
Explain how using antibacterial surface wipes could assist this process. **(5 marks)**

....

....

....

....

Score / 15

How well did you do?

0–7 Try again | 8–15 Getting there | 16–23 Good work | 24–31 Excellent!

For more information on this topic, see pages 66–67 of your Success Revision Guide.

Growth and Development

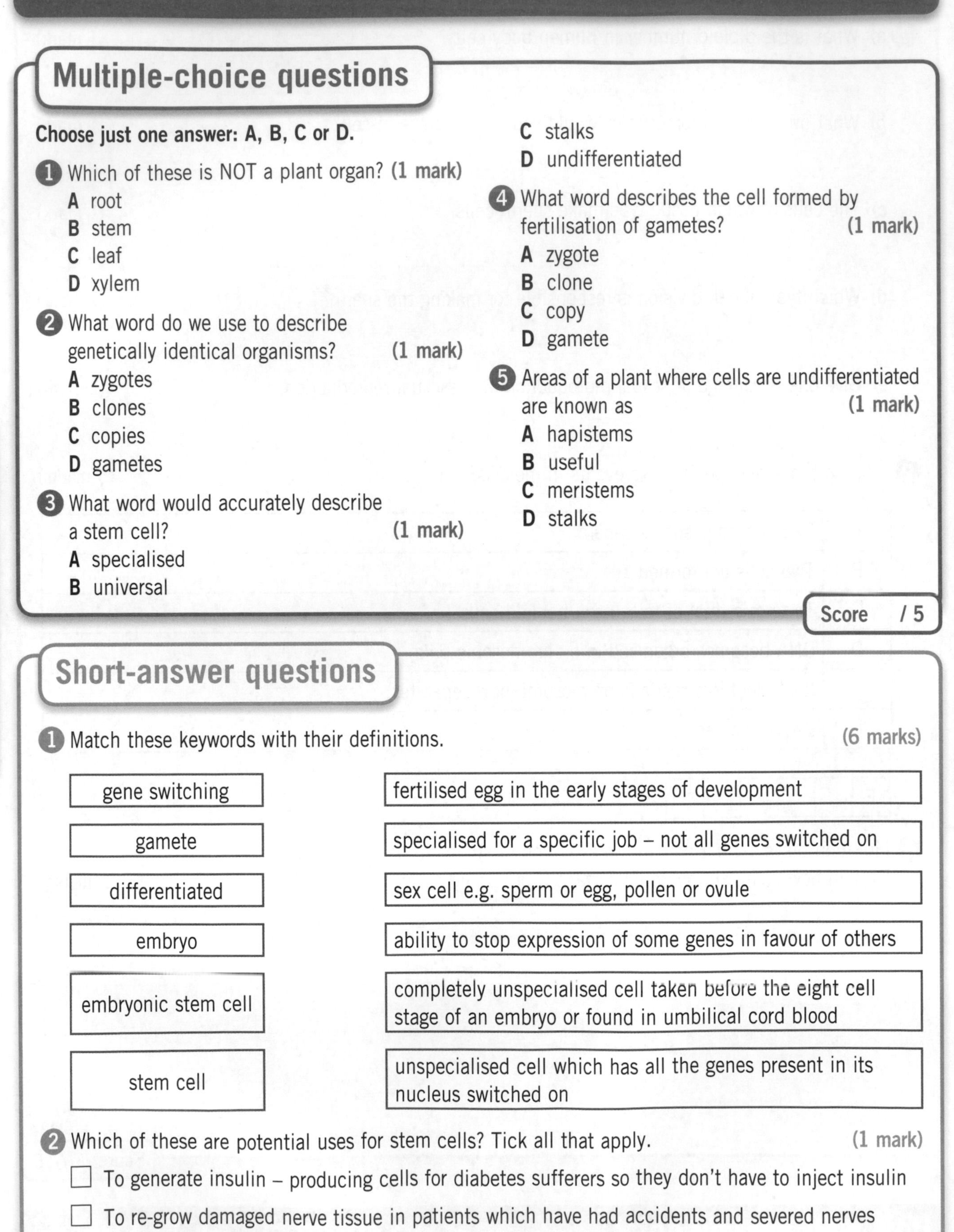

Multiple-choice questions

Choose just one answer: A, B, C or D.

1. Which of these is NOT a plant organ? **(1 mark)**
   - **A** root
   - **B** stem
   - **C** leaf
   - **D** xylem

2. What word do we use to describe genetically identical organisms? **(1 mark)**
   - **A** zygotes
   - **B** clones
   - **C** copies
   - **D** gametes

3. What word would accurately describe a stem cell? **(1 mark)**
   - **A** specialised
   - **B** universal
   - **C** stalks
   - **D** undifferentiated

4. What word describes the cell formed by fertilisation of gametes? **(1 mark)**
   - **A** zygote
   - **B** clone
   - **C** copy
   - **D** gamete

5. Areas of a plant where cells are undifferentiated are known as **(1 mark)**
   - **A** hapistems
   - **B** useful
   - **C** meristems
   - **D** stalks

Score / 5

Short-answer questions

1. Match these keywords with their definitions. **(6 marks)**

Keyword	Definition
gene switching	fertilised egg in the early stages of development
gamete	specialised for a specific job – not all genes switched on
differentiated	sex cell e.g. sperm or egg, pollen or ovule
embryo	ability to stop expression of some genes in favour of others
embryonic stem cell	completely unspecialised cell taken before the eight cell stage of an embryo or found in umbilical cord blood
stem cell	unspecialised cell which has all the genes present in its nucleus switched on

2. Which of these are potential uses for stem cells? Tick all that apply. **(1 mark)**
   - ☐ To generate insulin – producing cells for diabetes sufferers so they don't have to inject insulin
   - ☐ To re-grow damaged nerve tissue in patients which have had accidents and severed nerves
   - ☐ To grow new organs for transplant

Score / 7

GCSE-style questions

Answer all parts of all questions. Continue on a separate sheet of paper if necessary.

1 Look at the opinions about stem cell research.

Mark: Stem cell research should be banned. It is not right to interfere with nature like this.

Erica: Stem cell research is expensive. We can save more lives by investing in vaccines.

Javelle: Stem cells could potentially save and improve many lives by replacing ineffective or damaged tissue.

Ella: We need to look into the risks before we use cloned stem cells in humans.

a) Who has suggested a use for stem cells that might justify investment in research? **(1 mark)**

....

b) Who has made a comment based on ethical reasons? .... **(1 mark)**

c) Who is exercising the precautionary principle in their attitude to stem cell research? .... **(1 mark)**

2 a) Order these steps to producing measurements of dry mass for the mushrooms growing on a square metre of land. **(5 marks)**

A The mass of these mushrooms is measured.

B An area of 1 m^2 is chosen at random from a field (either by sampling with a quadrat or grid sampling).

C The mushrooms are removed from the oven and their mass is re-measured.

D The mushrooms in this area are picked.

E The mushrooms are placed in a warm oven for an hour.

F The mass is compared to the last measurement of mass taken.

G The process is repeated, returning the mushrooms to the oven and reweighing them alternately until no further loss of mass is recorded.

						G

b) What type of biomass is measured in step 'A' above? .... **(1 mark)**

c) What is the purpose of step G? .... **(1 mark)**

Score / 10

How well did you do?

0–5	Try again	6–11	Getting there	12–17	Good work	18–22	Excellent!

For more information on this topic, see pages 68–69 of your Success Revision Guide.

Transport in Cells

Multiple-choice questions

Choose just one answer: A, B, C or D.

1. Which of these is NOT the name of a method of transport in cells? **(1 mark)**
   - **A** active transport
   - **B** passive transport
   - **C** diffusion
   - **D** osmosis

2. Which of these words would NOT accurately describe diffusion? **(1 mark)**
   - **A** deliberate
   - **B** passive
   - **C** down a concentration gradient
   - **D** random

3. Which of these is a special case of diffusion relating to the movement of water? **(1 mark)**
   - **A** osmosis
   - **B** active transport
   - **C** flaccid
   - **D** turgid

4. Which of these describes a plant cell that has lost much of its cellular water, causing it to shrink? **(1 mark)**
   - **A** plasmolysed
   - **B** dry
   - **C** turgid
   - **D** arid

5. Which of these best describes a plant whose cells have lost a significant amount of water? **(1 mark)**
   - **A** turgid
   - **B** flaccid
   - **C** wilted
   - **D** arid

Score /5

Short-answer questions

1. Add these labels in the correct place on the diagram. (7 marks)

cell oxygen cell glucose carbon dioxide waste chemicals blood capillaries

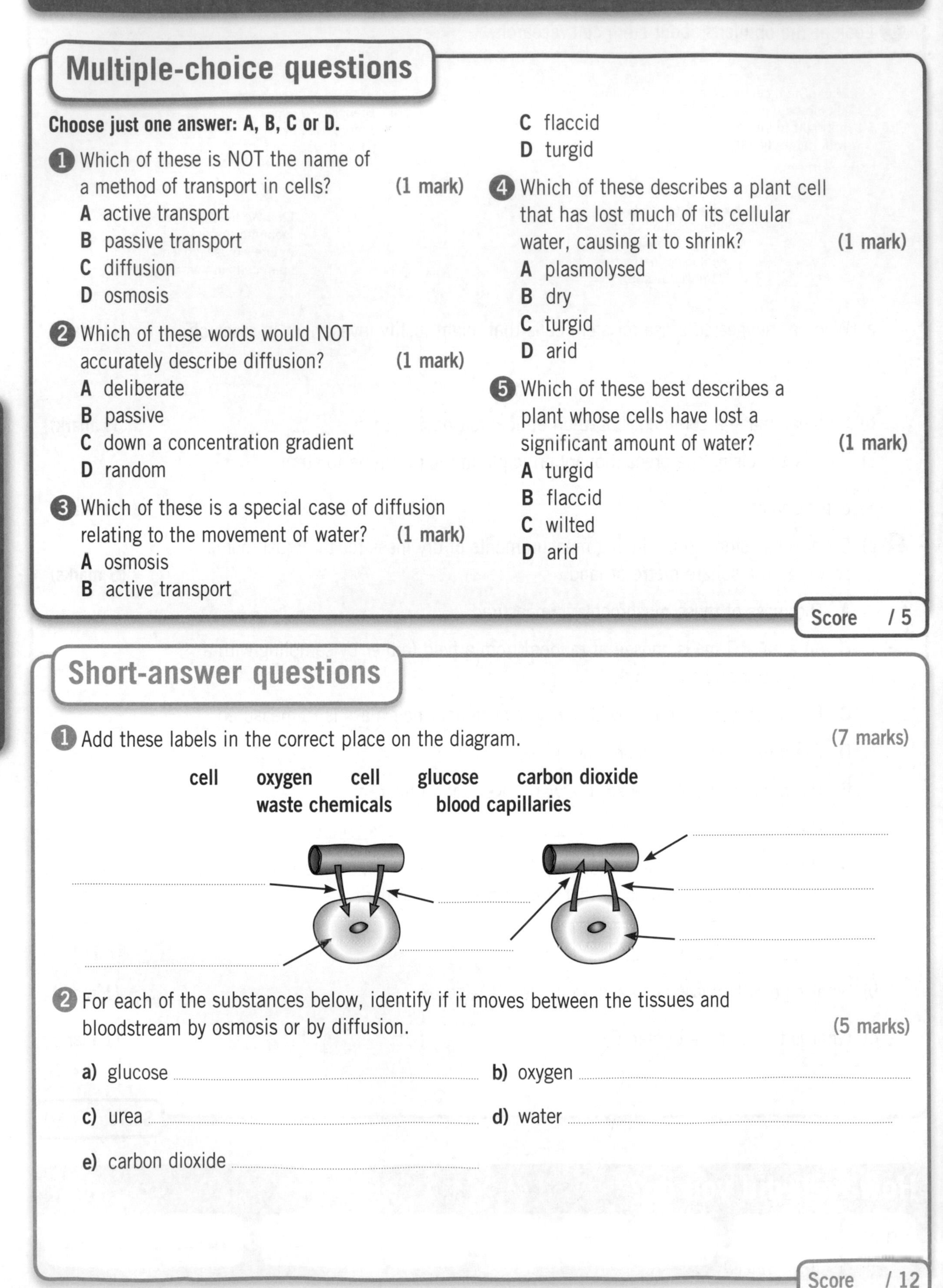

2. For each of the substances below, identify if it moves between the tissues and bloodstream by osmosis or by diffusion. (5 marks)

   a) glucose .... **b)** oxygen ....

   c) urea .... **d)** water ....

   e) carbon dioxide ....

Score /12

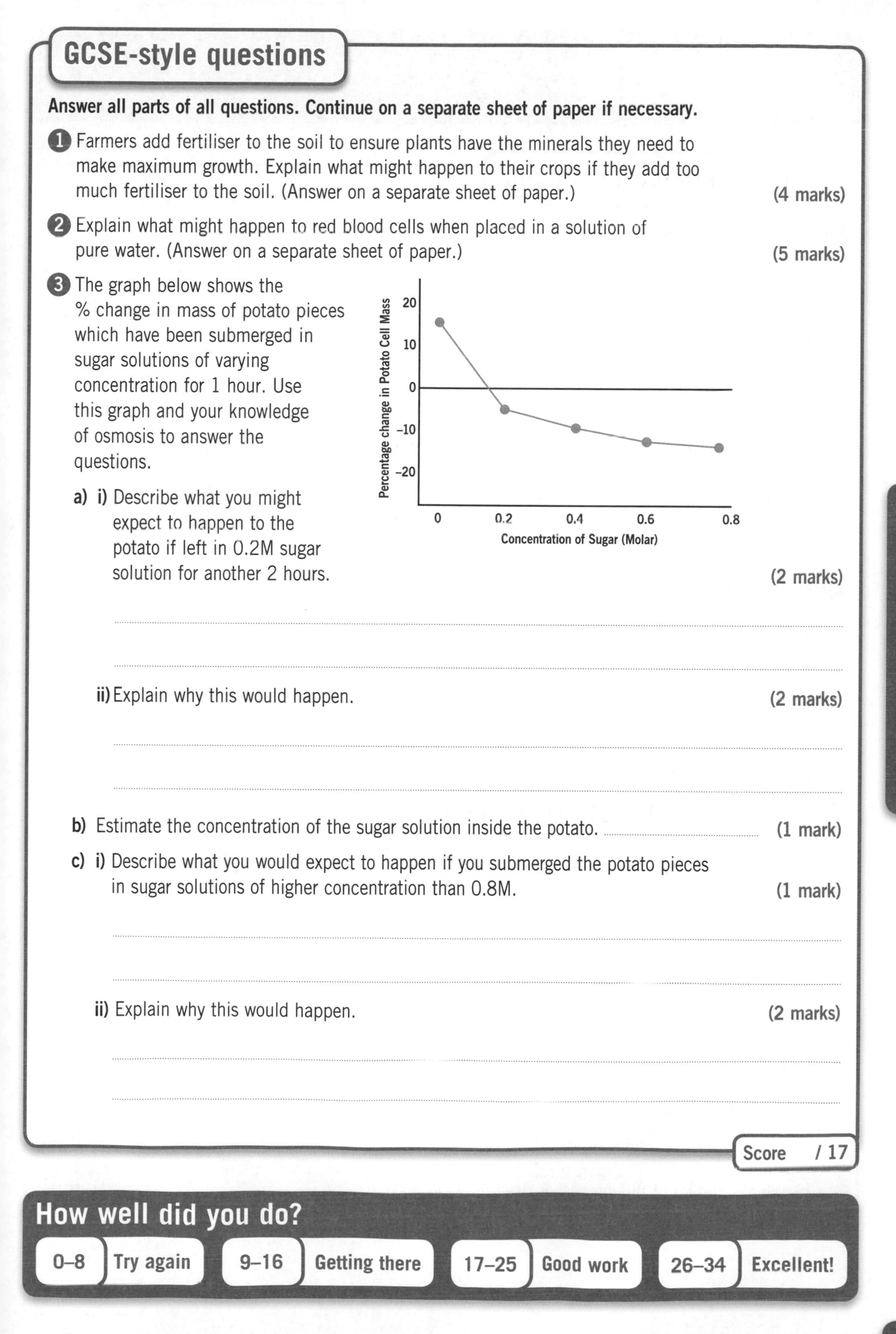

GCSE-style questions

Answer all parts of all questions. Continue on a separate sheet of paper if necessary.

1. Farmers add fertiliser to the soil to ensure plants have the minerals they need to make maximum growth. Explain what might happen to their crops if they add too much fertiliser to the soil. (Answer on a separate sheet of paper.) **(4 marks)**

2. Explain what might happen to red blood cells when placed in a solution of pure water. (Answer on a separate sheet of paper.) **(5 marks)**

3. The graph below shows the % change in mass of potato pieces which have been submerged in sugar solutions of varying concentration for 1 hour. Use this graph and your knowledge of osmosis to answer the questions.

a) i) Describe what you might expect to happen to the potato if left in 0.2M sugar solution for another 2 hours. **(2 marks)**

....

....

ii) Explain why this would happen. **(2 marks)**

....

....

b) Estimate the concentration of the sugar solution inside the potato. .... **(1 mark)**

c) i) Describe what you would expect to happen if you submerged the potato pieces in sugar solutions of higher concentration than 0.8M. **(1 mark)**

....

....

ii) Explain why this would happen. **(2 marks)**

....

....

Score / 17

How well did you do?

0–8 Try again | 9–16 Getting there | 17–25 Good work | 26–34 Excellent!

Cells and Organisation

For more information on this topic, see pages 70–71 of your Success Revision Guide.

Respiration

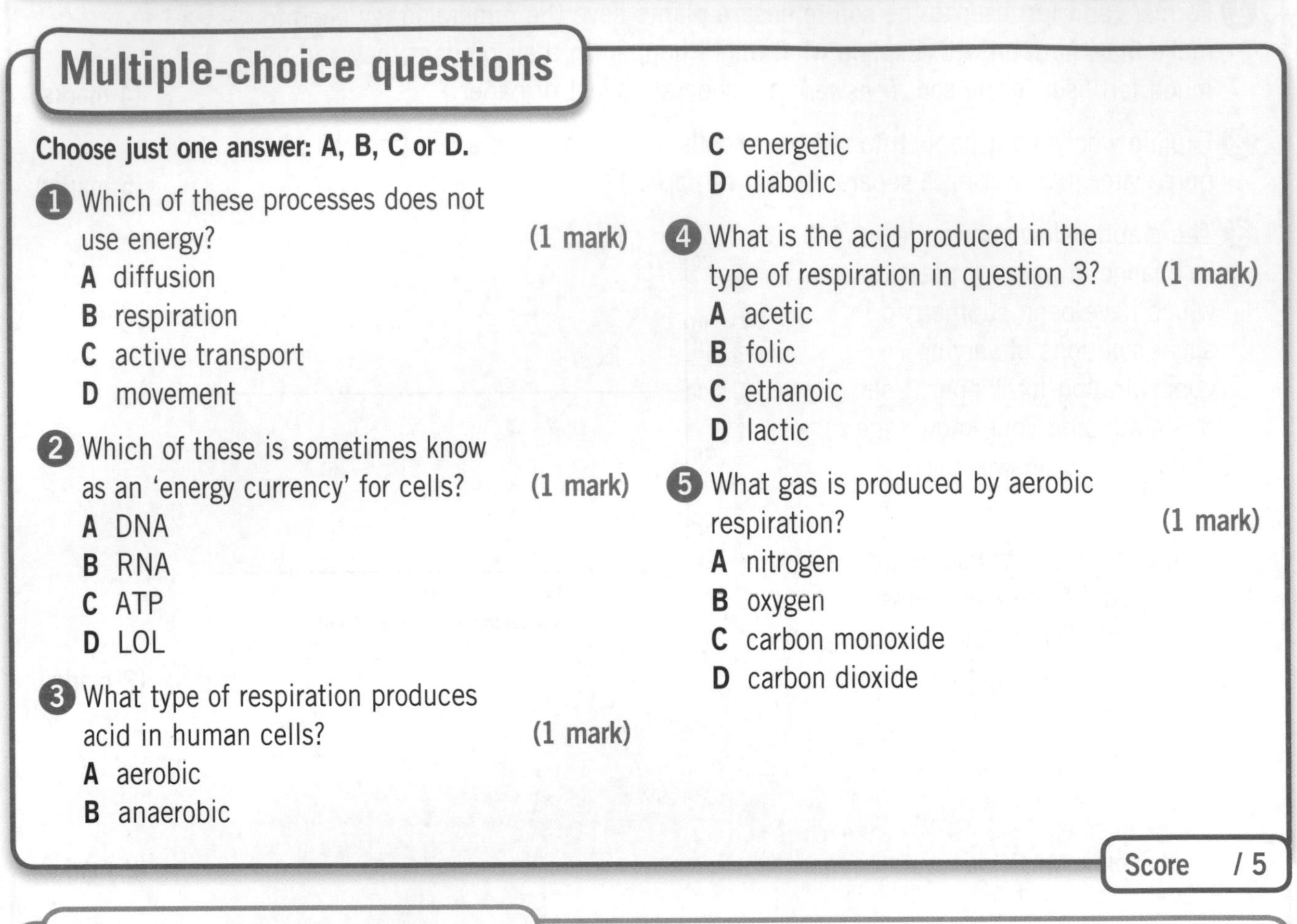

Multiple-choice questions

Choose just one answer: A, B, C or D.

1. Which of these processes does not use energy? **(1 mark)**
   - **A** diffusion
   - **B** respiration
   - **C** active transport
   - **D** movement
2. Which of these is sometimes know as an 'energy currency' for cells? **(1 mark)**
   - **A** DNA
   - **B** RNA
   - **C** ATP
   - **D** LOL
3. What type of respiration produces acid in human cells? **(1 mark)**
   - **A** aerobic
   - **B** anaerobic
   - **C** energetic
   - **D** diabolic
4. What is the acid produced in the type of respiration in question 3? **(1 mark)**
   - **A** acetic
   - **B** folic
   - **C** ethanoic
   - **D** lactic
5. What gas is produced by aerobic respiration? **(1 mark)**
   - **A** nitrogen
   - **B** oxygen
   - **C** carbon monoxide
   - **D** carbon dioxide

Score / 5

Short-answer questions

1. Complete the table, ticking (✓) all boxes that apply in each row. **(5 marks)**

Substance	Product of aerobic respiration	Product of anaerobic respiration
glucose		
carbon dioxide		
lactic acid		
energy		
water		
oxygen		

2. **a)** Anaerobic respiration in yeast is usefully used in industry. What is this process known as? .... **(1 mark)**

   b) Which product of this process is used in bread making? .... **(1 mark)**

   c) Which product of this process is used in beer making? .... **(1 mark)**

Score / 8

GCSE-style questions

Answer all parts of all questions. Continue on a separate sheet of paper if necessary.

1 Athletes completing a race continue to breathe heavily after the end of the race. What causes this to be necessary? (Answer on a separate sheet of paper.) **(6 marks)**

2 a) What warning signal does the body receive to prevent exercise from continuing if there is insufficient oxygen? .... **(1 mark)**

b) What problems would the creation of lactic acid pose for the body if levels of production were sustained? **(3 marks)**

....

....

3 What is the formula for calculating the respiratory quotient for an organism? **(1 mark)**

....

4 Which organelle of the cell is the site of respiration? .... **(1 mark)**

5 a) What is meant by the term metabolic rate? **(2 marks)**

....

b) Give three examples of activities that would increase the metabolic rate of an organism. **(3 marks)**

....

6

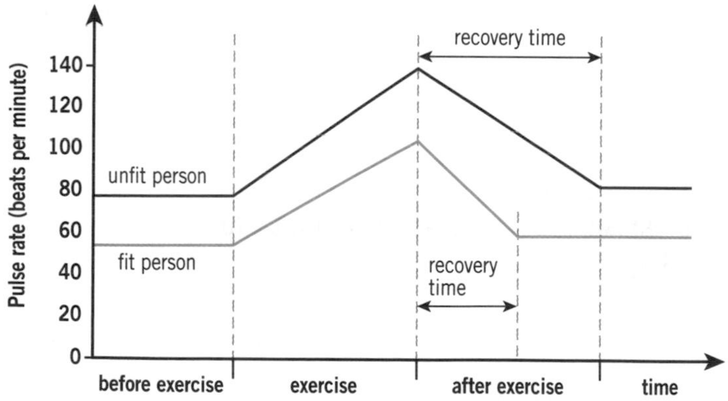

a) What do you notice about the pulse rate of an unfit person in comparison to a fit person during exercise? Explain why this is true. **(2 marks)**

....

b) What do you notice about the recovery time of the two individuals? Explain why. **(2 marks)**

....

Score / 21

How well did you do?

0–8	Try again	9–16	Getting there	17–25	Good work	26–34	Excellent!

For more information on this topic, see pages 72–73 of your Success Revision Guide.

Sampling Organisms

Multiple-choice questions

Choose just one answer: A, B, C or D.

1. Which word describes how estimates of population size could be obtained? **(1 mark)**
   A capturing
   B rationing
   C sampling
   D recapturing

2. What rectangular device is used to mark areas for observation? **(1 mark)**
   A quadrat
   B cuboid
   C box
   D transect

3. All the individuals of one species, in a particular area at a particular time are a **(1 mark)**
   A community
   B population
   C habitat
   D ecosystem

4. Which of these is unlikely to be used to capture a flying insect? **(1 mark)**
   A sticky paper
   B pitfall trap
   C UV light trap
   D net

5. Which of these describes a line of quadrats through a habitat? **(1 mark)**
   A transect
   B transport
   C transition
   D transistor

Score /5

Short-answer questions

1. Fill in the missing words in the paragraph. (9 marks)

captivity habitat organisms quadrat information
number sample pooters pitfall

Scientists are interested in making links between the .... that live in a .... and the physical features of the habitat. This .... can be used to help sustain populations of rare organisms in .... or to provide an understanding of the impact of human activity on the environment. It is not practical to count the .... of organisms in a population but instead an estimate based on a .... is used. Sampling plants can be done by measuring percentage cover or number of individuals in a .... placed randomly in an area. Samples of animals can be taken using .... and .... traps.

2. Give an example of a factor which might change and cause zonation along a seashore belt transect. (1 mark)

....

Score /10

GCSE-style questions

Answer all parts of all questions. Continue on a separate sheet of paper if necessary.

1 Some students have conducted a survey of their school garden. Use their notes to answer the questions that follow.

> We went out in the garden in our science lesson today. The weather was nice. We searched the whole garden and collected 10 snails. We put a blob of coloured paint on their shell and let them go again. The next day we went back and found 20 snails. Only 2 of the snails we found this time had paint on them. I didn't enjoy the lesson so much because the garden was wet. While we were in the garden we saw a trail of ants leading into a piece of dead wood. We could hear a woodpecker in the neighbouring garden but we didn't see it.

a) Give an example of a habitat mentioned in the notes. .... **(1 mark)**

b) Animal population size can be estimated by the capture-recapture technique. Use this formula to estimate how many snails there are living in the garden. **(2 marks)**

$$\text{Population size} = \frac{\text{number in 1st sample} \times \text{number in 2nd sample}}{\text{number in 2nd sample previously marked}}$$

Show your working.

....

....

....

c) Give one reason why the estimate may be higher than the actual number of snails in the garden. **(1 mark)**

....

d) Give one other reason why the estimate may be different from the actual number of snails in the garden. **(1 mark)**

....

e) If the students wanted to estimate the number of ants in the garden, why isn't capture–recapture a practical way of doing this? **(1 mark)**

....

2 **a)** What is meant by the term 'artificial ecosystem'? **(1 mark)**

....

b) Why might an 'artificial' ecosystem have less biodiversity than a 'natural' ecosystem? **(1 mark)**

....

....

Score /8

How well did you do?

1–6	Try again	7–12	Getting there	13–18	Good work	19–23	Excellent!

For more information on this topic, see pages 76–77 of your Success Revision Guide.

Photosynthesis

Multiple-choice questions

Choose just one answer: A, B, C, or D.

1. Which is the energy source for photosynthesis? **(1 mark)**
   A chemical
   B electrical
   C light
   D sound

2. Which of these is a raw material for photosynthesis? **(1 mark)**
   A carbon dioxide
   B oxygen
   C glucose
   D chemical energy

3. Which of these is NOT required for plant growth? **(1 mark)**
   A oxygen
   B carbon dioxide
   C minerals
   D soil

4. Which of these is NOT a rate limiting factor for photosynthesis? **(1 mark)**
   A light level
   B amount of carbon dioxide
   C temperature
   D amount of glucose

5. What is the name of the tissue in plant leaves responsible for the majority of photosynthesis? **(1 mark)**
   A spongy mesophyll layer
   B palisade layer
   C xylem
   D phloem

Score /5

Short-answer questions

1. Which structure in the plant cell is the site of photosynthesis? (1 mark)

....

2. Lack of magnesium in a plant results in stunted growth and yellow colouring of the plant. What pigment, vital in photosynthesis, is made using magnesium? (1 mark)

....

3. **True or false?** (4 marks)

	True	False
a) Increasing the temperature from 10°C to 20°C should nearly double the rate of photosynthesis.	☐	☐
b) Increasing the temperature from 40°C to 50°C should nearly double the rate of photosynthesis.	☐	☐
c) Adding more carbon dioxide will continually increase the amount of photosynthesis the plant can do.	☐	☐
d) Increasing the amount of light generally increases the rate of photosynthesis.	☐	☐

Score /6

GCSE-style questions

Answer all parts of all questions. Continue on a separate sheet of paper if necessary.

1 Examine the diagram of the cross section through a leaf.

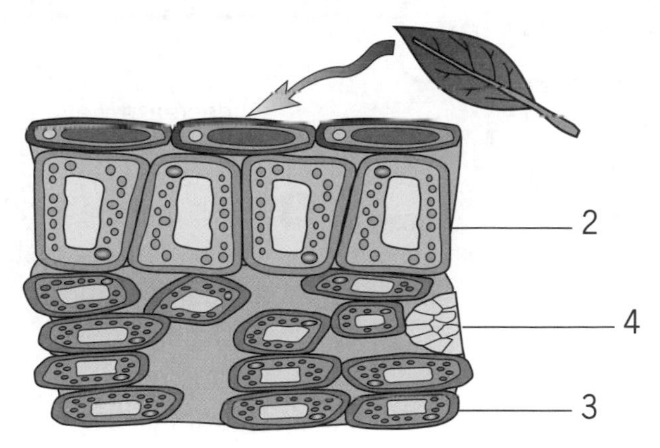

a) Identify the areas labelled 2–4 on the diagram and write them in the table below. **(3 marks)**

b) Describe the roles of these parts of leaf. **(3 marks)**

Number	Name	Role
	leaf vein (containing xylem and phloem)	
	palisade layer	
	spongy mesophyll layer	
1	underside of the leaf	Allows gases and water vapour to enter/ leave the leaf through small pores called stomata.

2 Sodium hydrogencarbonate indicator can be used to give an indication of the amount of carbon dioxide in a solution due to its change in colour at different pHs. The indicator changes in the following ways:

High levels of CO_2 = yellow	Atmospheric levels = red	Low levels of CO_2 = purple

In the following experimental conditions, what would you expect the colour of the indicator to be after 12 hours? Explain your reasons. (Answer on a separate sheet of paper.) **(3 marks)**

a) Open test tube containing indicator and pond water only.

b) Sealed test tube containing indicator, pond water and pond weed only, stored in a well-lit area.

c) Sealed test tube containing indicator, pond water and pond weed only, but covered in tin foil to prevent light entering.

Score /9

For more information on this topic, see pages 78–79 of your Success Revision Guide.

Food Production

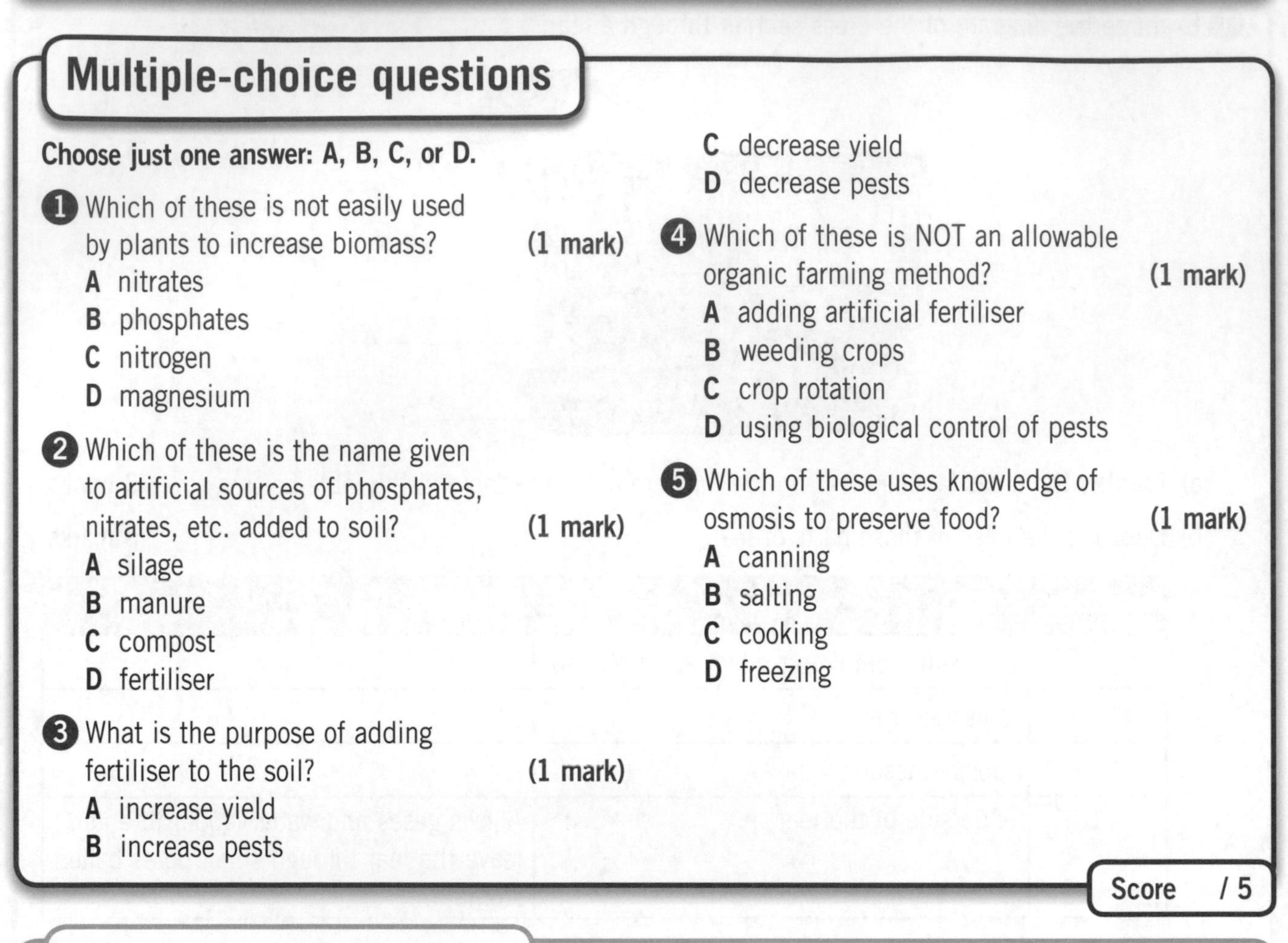

Multiple-choice questions

Choose just one answer: A, B, C, or D.

1. Which of these is not easily used by plants to increase biomass? **(1 mark)**
   A nitrates
   B phosphates
   C nitrogen
   D magnesium

2. Which of these is the name given to artificial sources of phosphates, nitrates, etc. added to soil? **(1 mark)**
   A silage
   B manure
   C compost
   D fertiliser

3. What is the purpose of adding fertiliser to the soil? **(1 mark)**
   A increase yield
   B increase pests
   C decrease yield
   D decrease pests

4. Which of these is NOT an allowable organic farming method? **(1 mark)**
   A adding artificial fertiliser
   B weeding crops
   C crop rotation
   D using biological control of pests

5. Which of these uses knowledge of osmosis to preserve food? **(1 mark)**
   A canning
   B salting
   C cooking
   D freezing

Score / 5

Short-answer questions

1. In each case, identify if the feature is related to intensive farming or organic farming. (7 marks)

Feature	Intensive? (✓)	Organic? (✓)
Few hedgerows		
Strips of land maintained 'unfarmed' to allow increased species diversity and encourage natural predators of pests to crops		
Biological control of pests		
Crop rotation including use of nitrogen-fixing crops		
Use of pesticides to reduce the number of weeds and increase crop yield		
Will use a higher amount of genetically modified crops to increase crop resistance to disease or pests for example		
Avoiding planting crops at times when the natural pests are active		

Score / 7

GCSE-style questions

Answer all parts of all questions. Continue on a separate sheet of paper if necessary.

1. Give the reasons why farmers might employ these methods:

a) using pesticides **(1 mark)**

....

b) using herbicides **(1 mark)**

....

c) vaccinating their livestock **(1 mark)**

....

....

d) keeping animals in confined areas **(1 mark)**

....

....

e) removing hedgerows .... **(1 mark)**

....

2. Explain how the following methods of food preservation reduce decay by bacteria. **(4 marks)**

Method	How it works
freezing	
cooking	
adding vinegar	
adding salt	

3. Irradiating food involves exposing it to high levels of radiation, which serves to cause fatal mutations in bacteria. This helps preserve the food. Can foods treated this way be labelled as organic in the EU? .... **(1 mark)**

4. Give one use for potassium in plants. **(1 mark)**

....

5. Give one use for phosphates in plants. **(1 mark)**

....

Score / 12

How well did you do?

0–6	Try again	7–12	Getting there	13–18	Good work	19–24	Excellent!

For more information on this topic, see pages 80–81 of your Success Revision Guide.

Transport in Animals

Multiple-choice questions

Choose just one answer: A, B, C or D.

1. Which vessels carry blood away from the heart? **(1 mark)**
   A arteries
   B veins
   C capillaries
   D lymph nodes

2. Which is the liquid part of the blood? **(1 mark)**
   A red blood cells
   B white blood cells
   C platelets
   D plasma

3. Which protein, used for carrying oxygen, is found in red blood cells? **(1 mark)**
   A haemoglobin
   B immunoglobin
   C fibrin
   D albumin

4. Which vessels carry blood at the highest pressure? **(1 mark)**
   A arteries
   B veins
   C capillaries
   D lymph nodes

5. Which of these vessels carries blood from the heart to the lungs? **(1 mark)**
   A vena cava
   B pulmonary vein
   C pulmonary artery
   D aorta

Score /5

Physiology

Short-answer questions

1. a) Add name labels to the four chambers of the heart and its four main blood vessels. (8 marks)

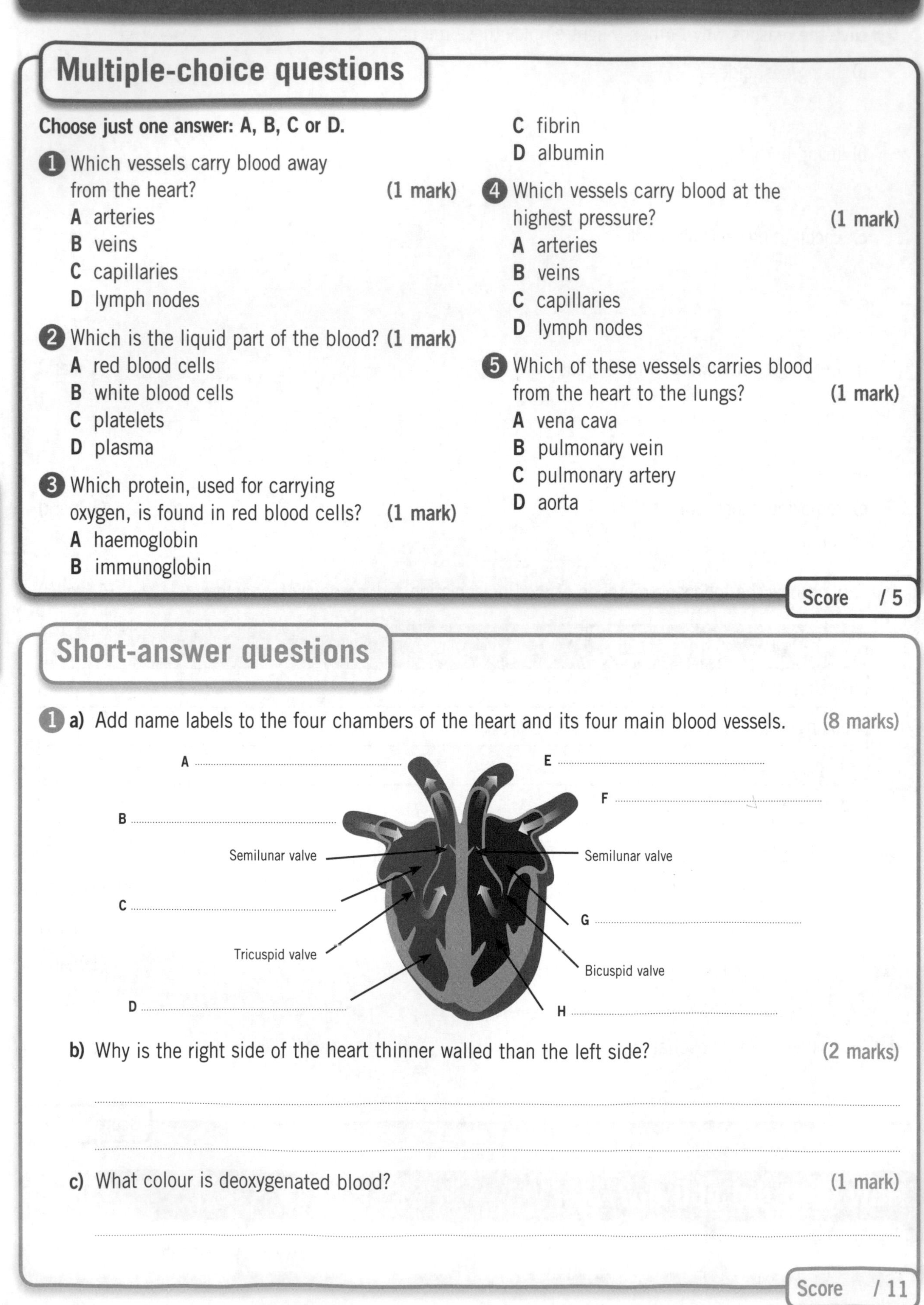

b) Why is the right side of the heart thinner walled than the left side? (2 marks)

....

....

c) What colour is deoxygenated blood? (1 mark)

....

Score /11

GCSE-style questions

Answer all parts of all questions. Continue on a separate sheet of paper if necessary.

1 What molecule is formed inside the red blood cell when it is carrying oxygen? **(1 mark)**

....

2

Vein

Thin wall

Lumen

Artery

Thick wall

Lumen

a) Identify a difference shown in the diagrams of the blood vessels above. **(1 mark)**

....

b) Explain why it is necessary for these vessels to differ in this way. **(2 marks)**

....

....

c) Give details of a further difference and its purpose. **(2 marks)**

....

....

....

3 a) In the table below, list three features of red blood cell physiology which make it good at carrying oxygen. **(3 marks)**

b) For each of your answers to a), explain how they benefit the red blood cell in its role. **(3 marks)**

Feature	Benefit in the role of carrying oxygen

Score / 12

How well did you do?

0–7 Try again | 8–14 Getting there | 15–21 Good work | 20–28 Excellent!

For more information on this topic, see pages 84–85 of your Success Revision Guide.

Transport in Plants

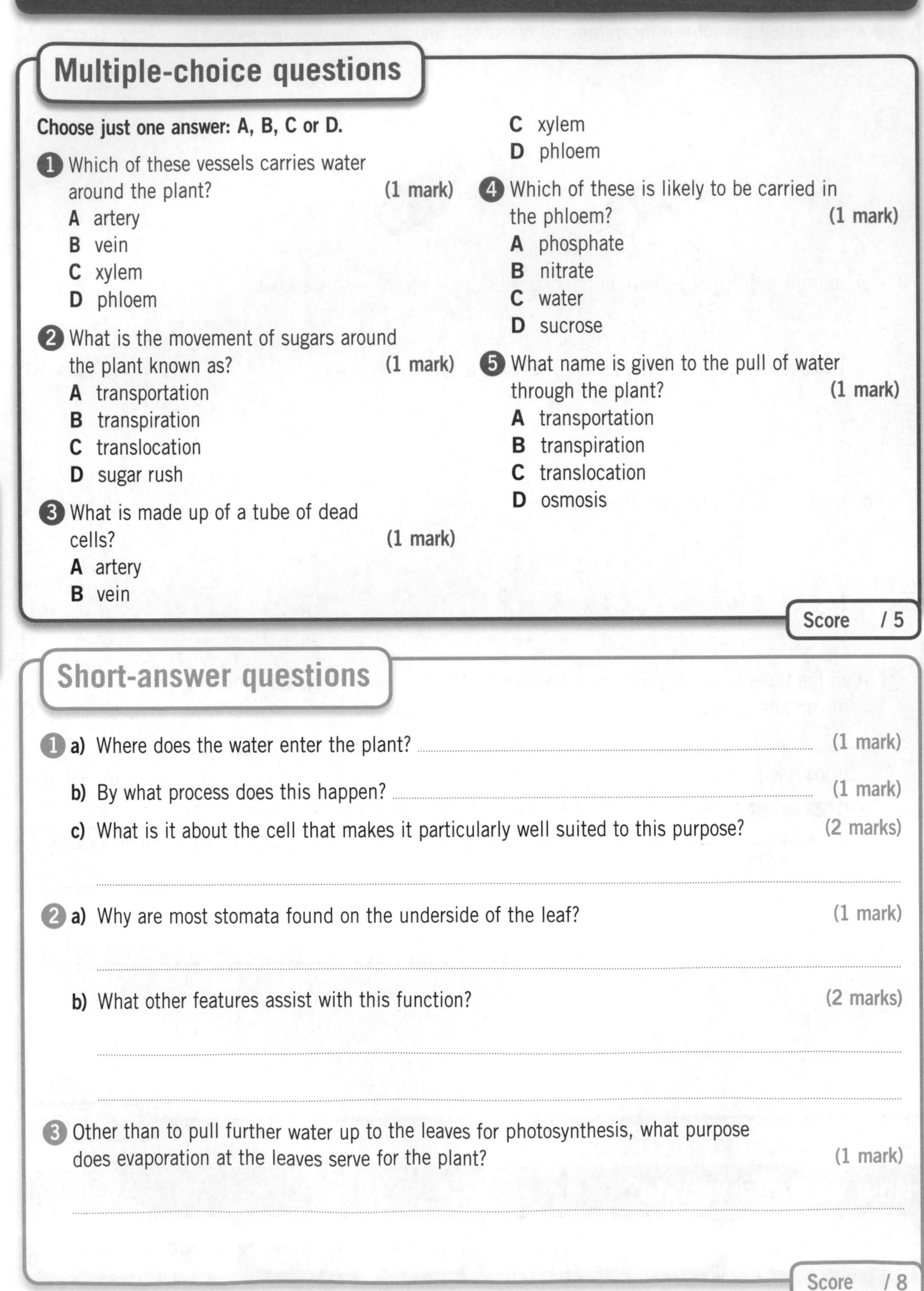

Multiple-choice questions

Choose just one answer: A, B, C or D.

1. Which of these vessels carries water around the plant? **(1 mark)**
   A artery
   B vein
   C xylem
   D phloem

2. What is the movement of sugars around the plant known as? **(1 mark)**
   A transportation
   B transpiration
   C translocation
   D sugar rush

3. What is made up of a tube of dead cells? **(1 mark)**
   A artery
   B vein
   C xylem
   D phloem

4. Which of these is likely to be carried in the phloem? **(1 mark)**
   A phosphate
   B nitrate
   C water
   D sucrose

5. What name is given to the pull of water through the plant? **(1 mark)**
   A transportation
   B transpiration
   C translocation
   D osmosis

Score / 5

Short-answer questions

1. **a)** Where does the water enter the plant? .... (1 mark)

   b) By what process does this happen? .... (1 mark)

   c) What is it about the cell that makes it particularly well suited to this purpose? (2 marks)

   ....

2. **a)** Why are most stomata found on the underside of the leaf? (1 mark)

   ....

   b) What other features assist with this function? (2 marks)

   ....

   ....

3. Other than to pull further water up to the leaves for photosynthesis, what purpose does evaporation at the leaves serve for the plant? (1 mark)

   ....

Score / 8

GCSE-style questions

Answer all parts of all questions. Continue on a separate sheet of paper if necessary.

1 When cut flowers are transported, air is introduced into the xylem vessels during periods when the flowers are out of water. This air breaks the transpiration stream. Explain why cutting the bottom off the stalks may prolong the life of the flowers once placed in a vase of water. **(6 marks)**

....

....

....

....

....

2 **a)** List four factors that affect the rate of transpiration. **(4 marks)**

....

....

b) For the answers above, explain how the transpiration rate is affected by the factor and the mechanism by which this happens. **(8 marks)**

....

....

....

....

....

....

c) Which would slow the transpiration rate down more: adding Vaseline to the upper surface or to the lower surface of the leaves? **(1 mark)**

....

d) Explain your answer. **(1 mark)**

....

Score / 20

How well did you do?

0–8	Try again	9–16	Getting there	17–25	Good work	26–33	Excellent!

Physiology

For more information on this topic, see pages 86–87 of your Success Revision Guide.

Digestion and Absorption

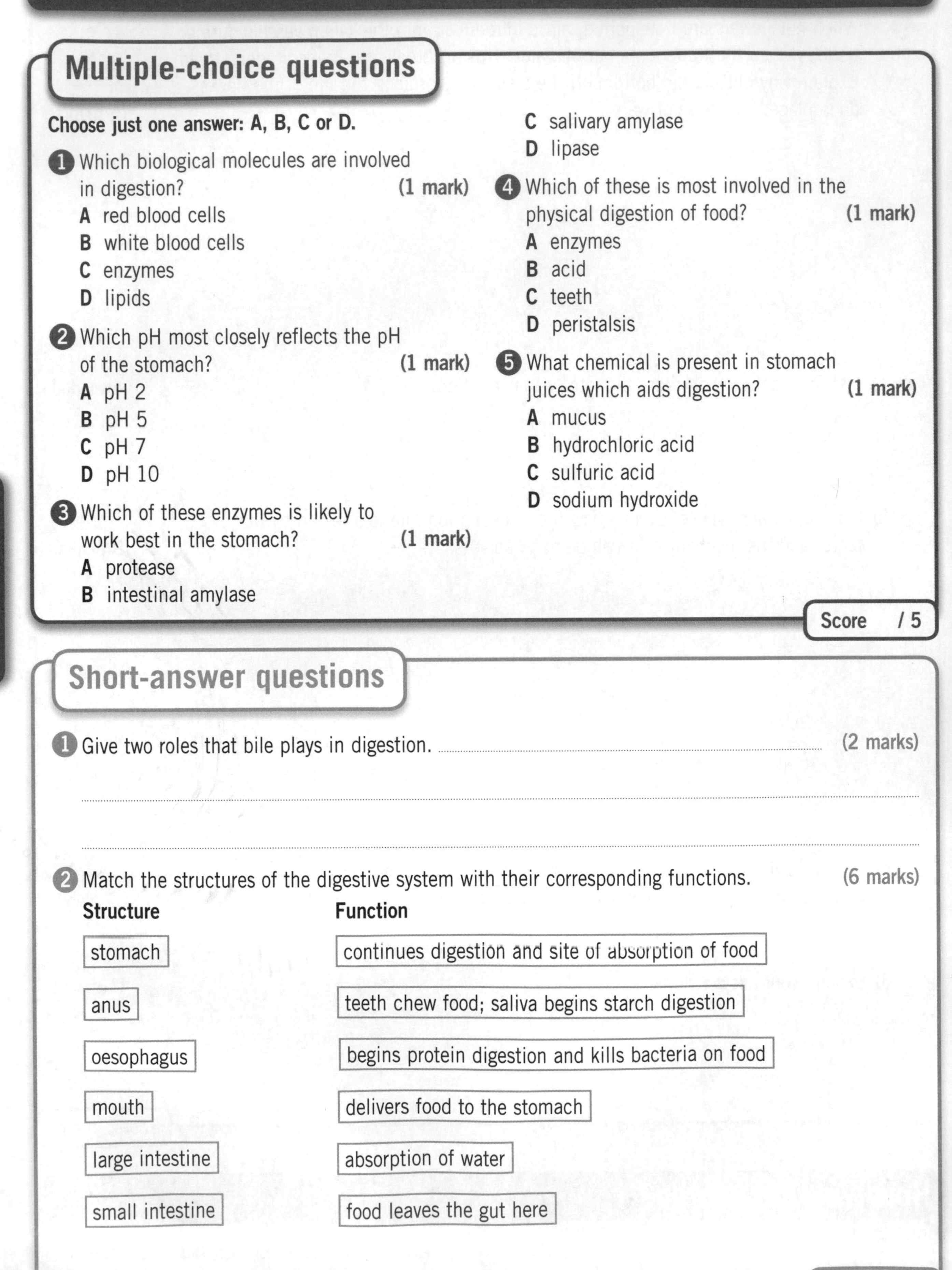

Multiple-choice questions

Choose just one answer: A, B, C or D.

1. Which biological molecules are involved in digestion? **(1 mark)**
   - **A** red blood cells
   - **B** white blood cells
   - **C** enzymes
   - **D** lipids
2. Which pH most closely reflects the pH of the stomach? **(1 mark)**
   - **A** pH 2
   - **B** pH 5
   - **C** pH 7
   - **D** pH 10
3. Which of these enzymes is likely to work best in the stomach? **(1 mark)**
   - **A** protease
   - **B** intestinal amylase
   - **C** salivary amylase
   - **D** lipase
4. Which of these is most involved in the physical digestion of food? **(1 mark)**
   - **A** enzymes
   - **B** acid
   - **C** teeth
   - **D** peristalsis
5. What chemical is present in stomach juices which aids digestion? **(1 mark)**
   - **A** mucus
   - **B** hydrochloric acid
   - **C** sulfuric acid
   - **D** sodium hydroxide

Score /5

Short-answer questions

1. Give two roles that bile plays in digestion. .... (2 marks)

   ....

   ....

2. Match the structures of the digestive system with their corresponding functions. (6 marks)

Structure	Function
stomach	continues digestion and site of absorption of food
anus	teeth chew food; saliva begins starch digestion
oesophagus	begins protein digestion and kills bacteria on food
mouth	delivers food to the stomach
large intestine	absorption of water
small intestine	food leaves the gut here

Score /8

GCSE-style questions

Answer all parts of all questions. Continue on a separate sheet of paper if necessary.

1. What is the benefit of emulsifying fats in terms of lipase action? **(1 mark)**

2. **a)** What smaller molecules is starch digested into? **(1 mark)**

   b) Why is digestion of food necessary? **(2 marks)**

3. Describe two features of the small intestine which increase the level of absorption of food. **(2 marks)**

4. By what process does food enter the blood stream? **(1 mark)**

5. Into what type of blood vessel would food move from the small intestine? **(1 mark)**

6. **a)** Food entering the blood system at the small intestine is destined for which organ? **(1 mark)**

   b) In which blood vessel will it travel to get there? **(1 mark)**

   c) What waste product, produced from the excess protein, is released into the blood stream? **(1 mark)**

7. Put these parts of the digestive system in the order they are met by food passing through them. **(6 marks)**

   oesophagus mouth small intestine
   anus stomach large intestine

Score / 17

How well did you do?

0–7	Try again	8–15	Getting there	16–22	Good work	23–30	Excellent!

For more information on this topic, see pages 88–89 of your Success Revision Guide.

The Heart and Circulation

Multiple-choice questions

Choose just one answer: A, B, C or D.

1. Which organisms have an open circulatory system? **(1 mark)**
   A fish
   B birds
   C insects
   D mammals

2. Which hormone alters the heart rate? **(1 mark)**
   A testosterone
   B insulin
   C glucagon
   D adrenaline

3. An inherited condition in which the blood does not clot is called **(1 mark)**
   A haemophilia
   B sickle cell anaemia
   C rhesus factor
   D cystic fibrosis

4. Which of these fruits affects blood clotting? **(1 mark)**
   A apple
   B blueberry
   C cranberry
   D pear

5. Which blood group produces antibodies to A and B antigens? **(1 mark)**
   A O
   B A
   C B
   D AB

Score / 5

Short-answer questions

1. Outline the sequence of events in the cardiac cycle. **(4 marks)**

   ....

   ....

   ....

2. Blood group is controlled by genes. What are the two blood grouping systems called? **(2 marks)**

   ....

3. Complete the following. **(5 marks)**

   The rate of the heart is controlled by the .... This is divided into two sections, the SAN and the .... An electrical current is produced, which spreads through the heart causing it to .... and hormones can alter the heart rate. It is possible to diagnose abnormalities in the heart's electric current by using an .... machine.

Score / 11

Use, Damage and Repair

GCSE-style questions

Answer all parts of all questions. Continue on a separate sheet of paper if necessary.

1. Compare the following terms, giving an example for each. **(4 marks)**

   a) Open and closed circulatory system

   ....

   ....

   b) Single and double circulatory system **(4 marks)**

   ....

   ....

2. An investigation is being carried out into the effect of exercise on the heart rate. Describe how the heart rate can be measured. **(2 marks)**

   ....

   ....

   ....

3. The table below shows different people's blood groups. Complete the table for the antibodies found. **(4 marks)**

Blood group	Contains antibodies against
A	
B	
AB	
O	

4. **a)** What component of blood causes blood to clot? .... **(1 mark)**

   b) Explain how a scab forms when a person cuts their skin. **(4 marks)**

   ....

   ....

   ....

   ....

   c) Name a drug that will reduce blood clotting. .... **(1 mark)**

Score / 20

How well did you do?

0–9	Try again	10–19	Getting there	20–28	Good work	29–36	Excellent!

For more information on this topic, see pages 92–93 of your Success Revision Guide.

The Skeleton and Exercise

Multiple-choice questions

Choose just one answer: A, B, C or D.

1. Which of the following is a ball and socket joint? **(1 mark)**
   - **A** elbow
   - **B** knee
   - **C** hip
   - **D** neck

2. Muscles are attached to bone by which of the following? **(1 mark)**
   - **A** tendons
   - **B** ligaments
   - **C** cartilage
   - **D** bonds

3. What is the purpose of synovial joints? **(1 mark)**
   - **A** allow movement in all directions
   - **B** reduce friction
   - **C** provide strength
   - **D** protect organs

4. Which of the following is an invertebrate? **(1 mark)**
   - **A** frog
   - **B** snake
   - **C** beetle
   - **D** horse

5. Anatagonistic muscles work by which of the following? **(1 mark)**
   - **A** pulling in one direction
   - **B** pulling in opposite directions
   - **C** pushing in opposite directions
   - **D** pulling and pushing

Score / 5

Short-answer questions

1. The diagram shows a synovial joint. Label structures A, B and C and describe their functions. **(6 marks)**

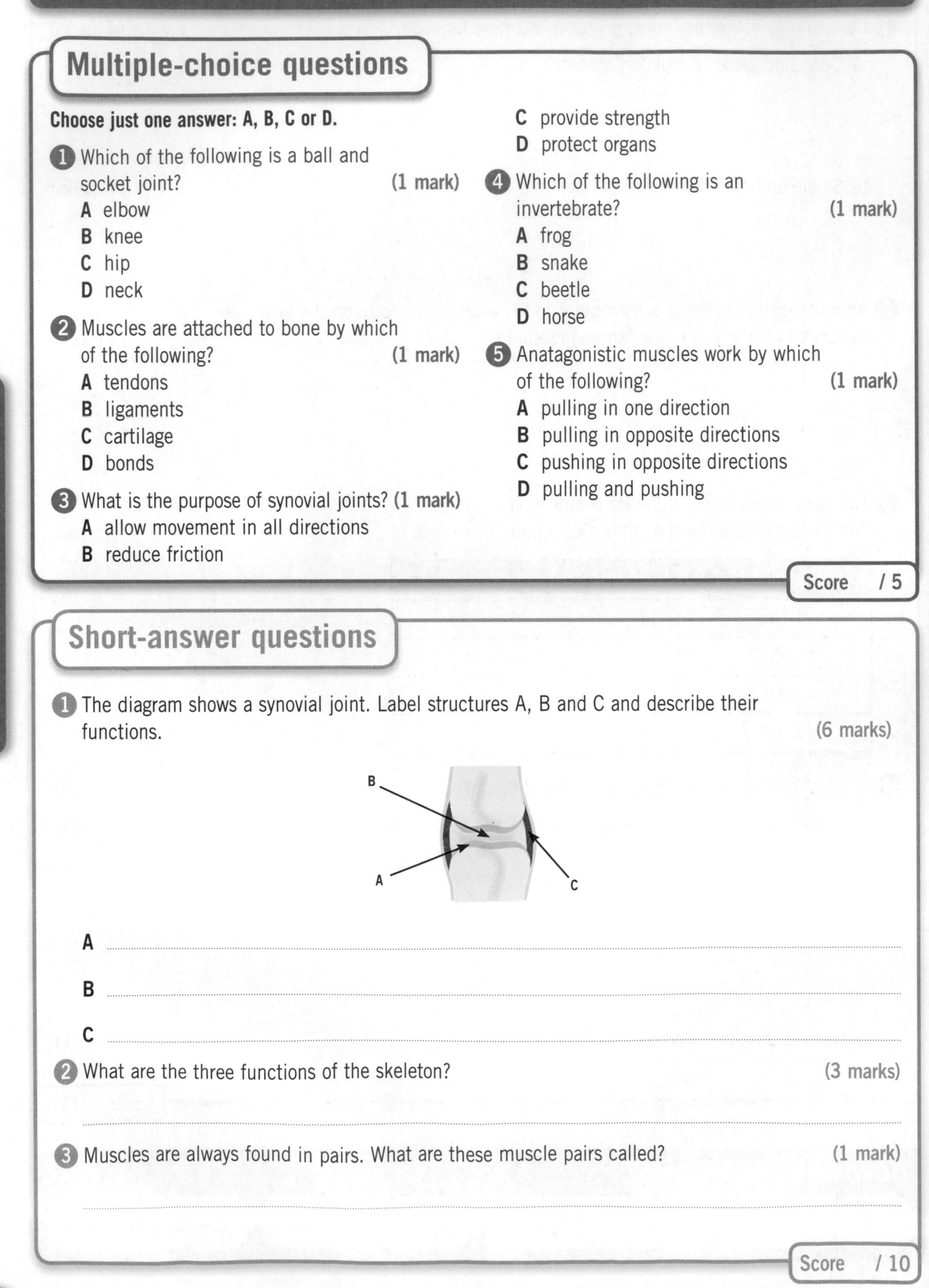

A ....

B ....

C ....

2. What are the three functions of the skeleton? **(3 marks)**

....

3. Muscles are always found in pairs. What are these muscle pairs called? **(1 mark)**

....

Score / 10

GCSE-style questions

Answer all parts of all questions. Continue on a separate sheet of paper if necessary.

1 The diagram shows the bones and muscles in the arm.

a) Which muscles control the movement of the lower arm? **(2 marks)**

....

b) Explain how these work to move the arm down. **(3 marks)**

....

....

c) Which type of synovial joint is found at the elbow? **(1 mark)**

....

2 **a)** When a person embarks on a fitness regime what factors in their life should they consider? **(3 marks)**

....

....

b) During exercise, what can the person monitor to help assess their progress? **(2 marks)**

....

....

3 There are two types of skeleton, external and internal. Describe the advantages and disadvantages of both. **(4 marks)**

External ....

Internal ....

Score / 15

How well did you do?

0–7	Try again	8–15	Getting there	16–22	Good work	23–30	Excellent!

For more information on this topic, see pages 94–95 of your Success Revision Guide.

The Excretory System

Multiple-choice questions

Choose just one answer: A, B, C or D.

1. Which blood vessel supplies the kidney? **(1 mark)**
   - **A** aorta
   - **B** renal artery
   - **C** hepatic artery
   - **D** vena cava

2. Where does ultra filtration occur? **(1 mark)**
   - **A** glomerulus
   - **B** collecting duct
   - **C** loop of Henle
   - **D** convoluted tubule

3. Which toxic substance is filtered by the kidney? **(1 mark)**
   - **A** salt
   - **B** water
   - **C** amino acids
   - **D** urea

4. Which organ breaks down excess amino acids? **(1 mark)**
   - **A** liver
   - **B** kidney
   - **C** stomach
   - **D** intestine

5. Which of the following is not an excretory product? **(1 mark)**
   - **A** carbon dioxide
   - **B** urine
   - **C** faeces
   - **D** water

Score /5

Short-answer questions

1. Label the diagram of the nephron. (5 marks)

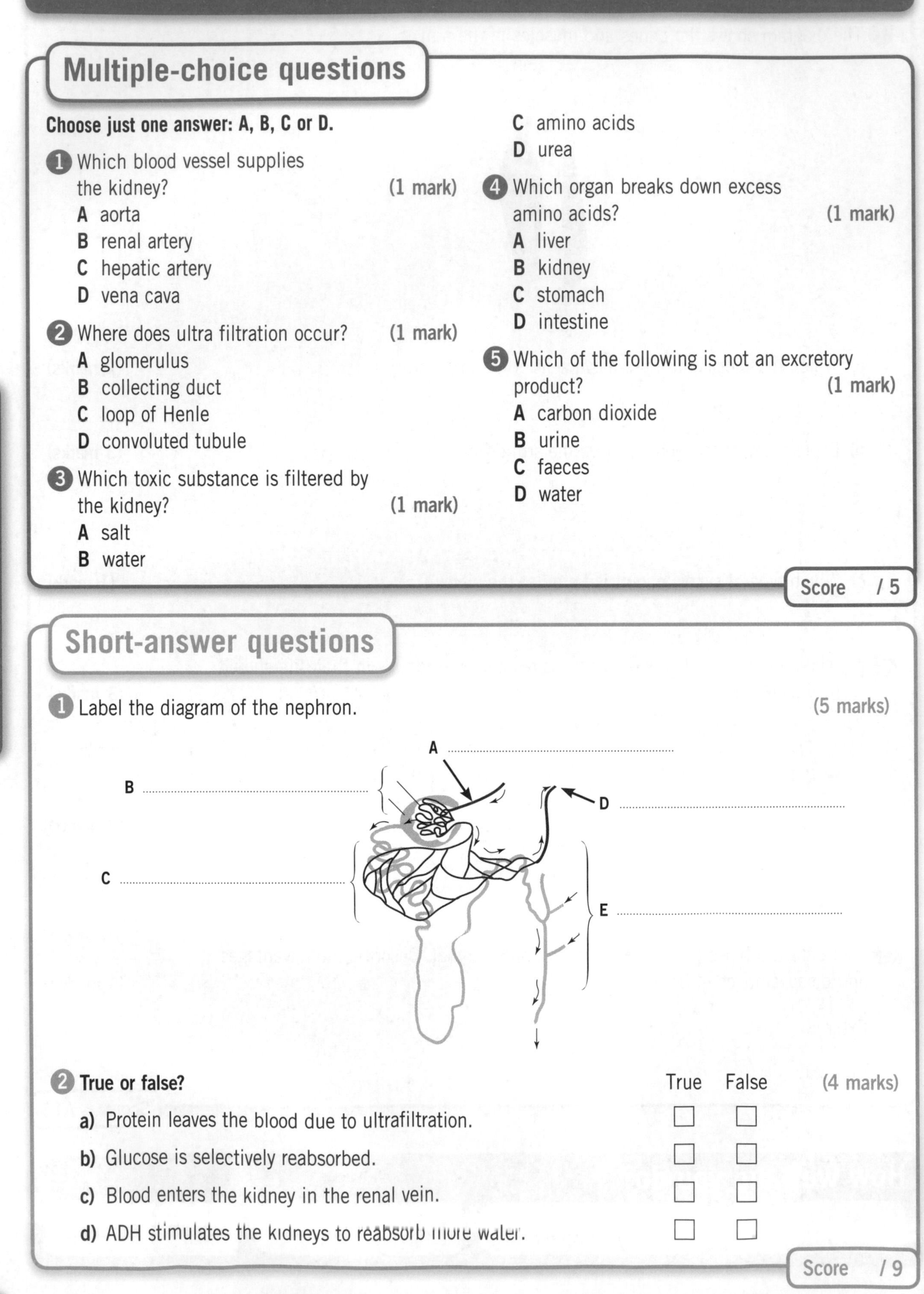

2. **True or false?** (4 marks)

	True	False
a) Protein leaves the blood due to ultrafiltration.	☐	☐
b) Glucose is selectively reabsorbed.	☐	☐
c) Blood enters the kidney in the renal vein.	☐	☐
d) ADH stimulates the kidneys to reabsorb more water.	☐	☐

Score /9

GCSE-style questions

Answer all parts of all questions. Continue on a separate sheet of paper if necessary.

1. Compare the terms excretion and egestion. **(2 marks)**

....

....

....

2. Describe the processes that occur within the kidney to maintain the desired concentration of glucose, water and urine. **(6 marks)**

....

....

....

....

....

....

....

....

3. **a)** Urea is produced from which chemical food molecule? .... **(1 mark)**

   b) Where is urea produced? .... **(1 mark)**

4. Complete the table indicating if the molecule is present (✓) or absent (✗). **(3 marks)**

Molecule	Blood in renal artery	Filtrate in capsule	Urine
protein			
glucose			
urea			

Score / 13

How well did you do?

0–6 Try again | 7–12 Getting there | 13–19 Good work | 20–27 Excellent!

For more information on this topic, see pages 96–97 of your Success Revision Guide.

Breathing

Multiple-choice questions

Choose just one answer: A, B, C or D.

1. In which organism does gaseous exchange involve alveoli? **(1 mark)**
   - **A** worm
   - **B** fish
   - **C** frog
   - **D** human

2. Which of the following does NOT aid gaseous exchange? **(1 mark)**
   - **A** small surface area
   - **B** thin membrane
   - **C** permeable
   - **D** rich blood supply

3. Which of the following occurs in inhalation? **(1 mark)**
   - **A** intercostal muscles relax
   - **B** diaphragm relaxes
   - **C** ribs move up and outwards
   - **D** pressure in thorax increases

4. What do fish use for gas exchange? **(1 mark)**
   - **A** gills
   - **B** lungs
   - **C** alveoli
   - **D** skin

5. Which best describes respiration? **(1 mark)**
   - **A** ventilation of the lungs
   - **B** oxygen diffusing into the blood
   - **C** exchange of gases at the lungs
   - **D** reaction producing energy

Score /5

Short-answer questions

1. Describe the following: **(3 marks)**

   Breathing ....

   Gaseous exchange ....

   Respiration ....

2. **True or false?** **(6 marks)**

	True	False
a) Adult frogs have lungs to aid gaseous exchange.	☐	☐
b) Large active organisms can rely on diffusion over their body surface to gain oxygen.	☐	☐
c) Spirometers measure the volume of air exchanged during breathing.	☐	☐
d) Vital capacity is the minimum air lungs can pass in or out per breath.	☐	☐
e) Residual volume is the air in the lungs during normal breathing.	☐	☐
f) Filaments in fish gills have a large surface area.	☐	☐

Score /9

GCSE-style questions

Answer all parts of all questions. Continue on a separate sheet of paper if necessary.

1. Explain why a single-celled organism, e.g. bacteria, can obtain its oxygen requirements by diffusion but large organisms, e.g. humans, require special respiratory surfaces. **(2 marks)**

   ....

   ....

2. Complete the table by crossing through the incorrect word. **(5 marks)**

Structure	Inhalation (breathing in)
intercostal muscles	contract / relax
diaphragm	contract / relax
ribcage moves	up and out / down and in
volume inside pleural cavity	increases / decreases
pressure inside pleural cavity	increases / decreases

3. An individual's breathing is monitored using a spirometer. Below is the trace that is produced.

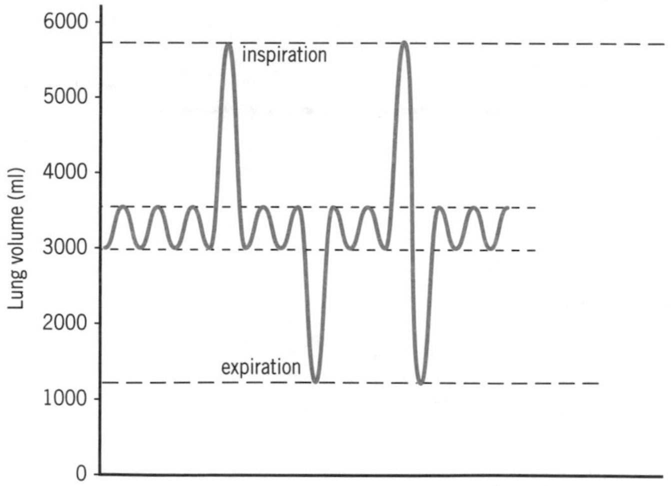

Lung volume (ml)
6000
5000
4000
3000
2000
1000
0
inspiration
expiration

   a) Calculate the tidal volume for this person. **(1 mark)**

   ....

   ....

   b) Which values would you need to add together to calculate the total lung capacity? **(2 marks)**

   ....

Score /10

How well did you do?

0–6	Try again	7–12	Getting there	13–18	Good work	19–24	Excellent!

For more information on this topic, see pages 98–99 of your Success Revision Guide.

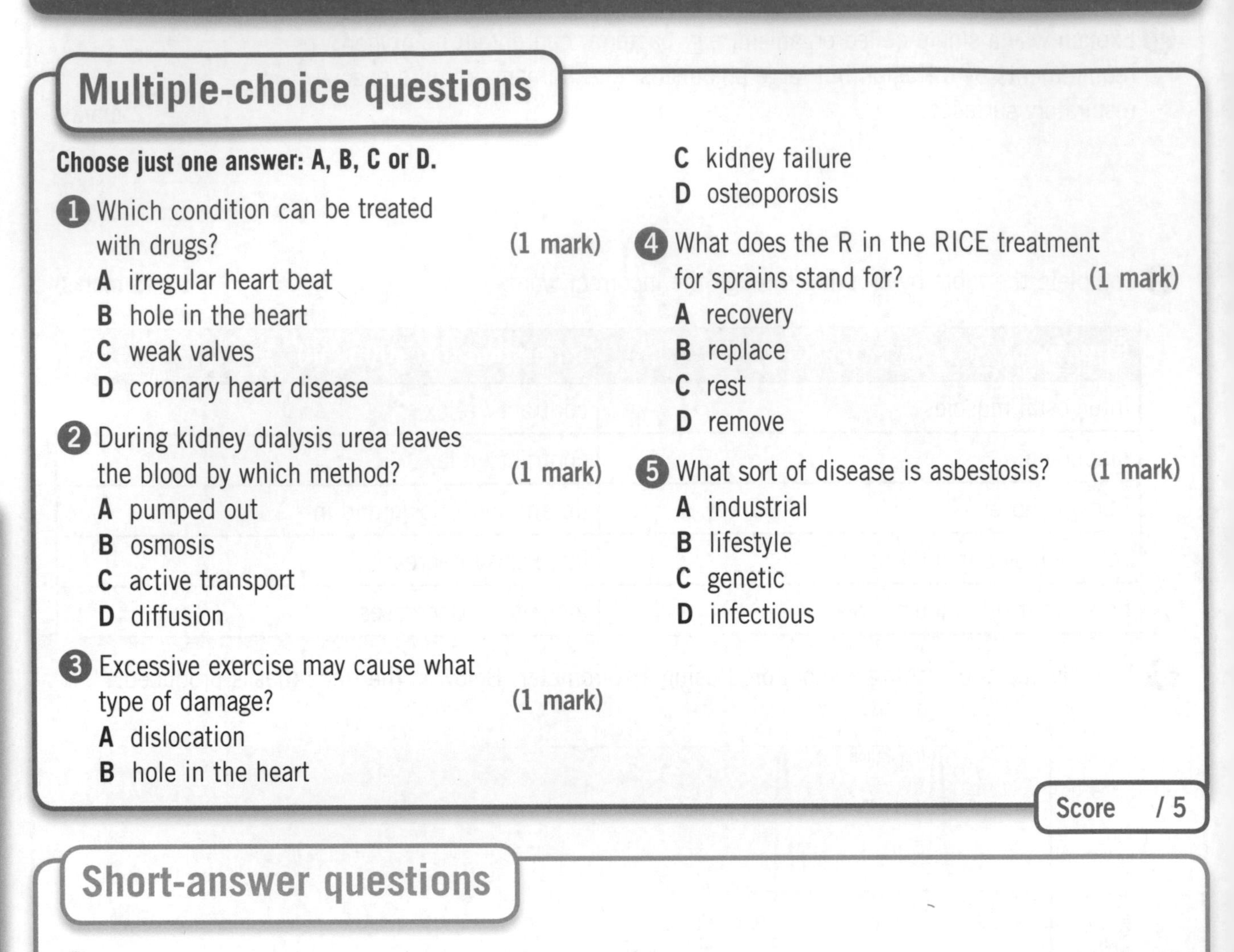

Damage and Repair

Multiple-choice questions

Choose just one answer: A, B, C or D.

1. Which condition can be treated with drugs? **(1 mark)**
   - **A** irregular heart beat
   - **B** hole in the heart
   - **C** weak valves
   - **D** coronary heart disease
2. During kidney dialysis urea leaves the blood by which method? **(1 mark)**
   - **A** pumped out
   - **B** osmosis
   - **C** active transport
   - **D** diffusion
3. Excessive exercise may cause what type of damage? **(1 mark)**
   - **A** dislocation
   - **B** hole in the heart
   - **C** kidney failure
   - **D** osteoporosis
4. What does the R in the RICE treatment for sprains stand for? **(1 mark)**
   - **A** recovery
   - **B** replace
   - **C** rest
   - **D** remove
5. What sort of disease is asbestosis? **(1 mark)**
   - **A** industrial
   - **B** lifestyle
   - **C** genetic
   - **D** infectious

Score /5

Short-answer questions

1. What are the treatments available if kidney failure occurs? (2 marks)

   ....

2. Complete the table for heart conditions and diseases. (4 marks)

Condition	Effect on the body	Treatment
irregular heart beat	less oxygen in the blood	
	blood in the right and left side of the heart can mix	surgery
weak valves	reduces blood circulation	
coronary heart disease		by-pass surgery

3. Name three types of bone fracture. (3 marks)

   ....

Score /9

GCSE-style questions

Answer all parts of all questions. Continue on a separate sheet of paper if necessary.

1 The diagram shows a kidney dialysis machine. Kidney dialysis removes urea from the blood because the person's kidneys have failed.

a) Why does the dialysis fluid contain glucose, amino acids and some salts? **(2 marks)**

....

b) By which process does urea leave the blood and enter the dialysis fluid? **(1 mark)**

....

c) Why is blood removed from a vein in the arm and not an artery? **(1 mark)**

....

d) Why would it be an advantage if the partially permeable membrane was folded instead of flat? **(2 marks)**

....

2 **a)** Sports physiotherapists often deal with dislocations, sprains and tendon damage. Why do these types of injuries often occur? **(1 mark)**

....

b) How would a sprain be treated? **(4 marks)**

....

c) Elderly people are more likely to suffer from bone fractures. Explain why this is the case. **(2 marks)**

....

3 **a)** Describe the symptoms of asthma. **(2 marks)**

....

b) What happens to the respiratory system during an asthma attack? **(3 marks)**

....

c) How is asthma treated? .... **(1 mark)**

Score / 19

How well did you do?

0–8	Try again	9–16	Getting there	17–25	Good work	26–33	Excellent!

For more information on this topic, see pages 100–101 of your Success Revision Guide.

Transplants and Donations

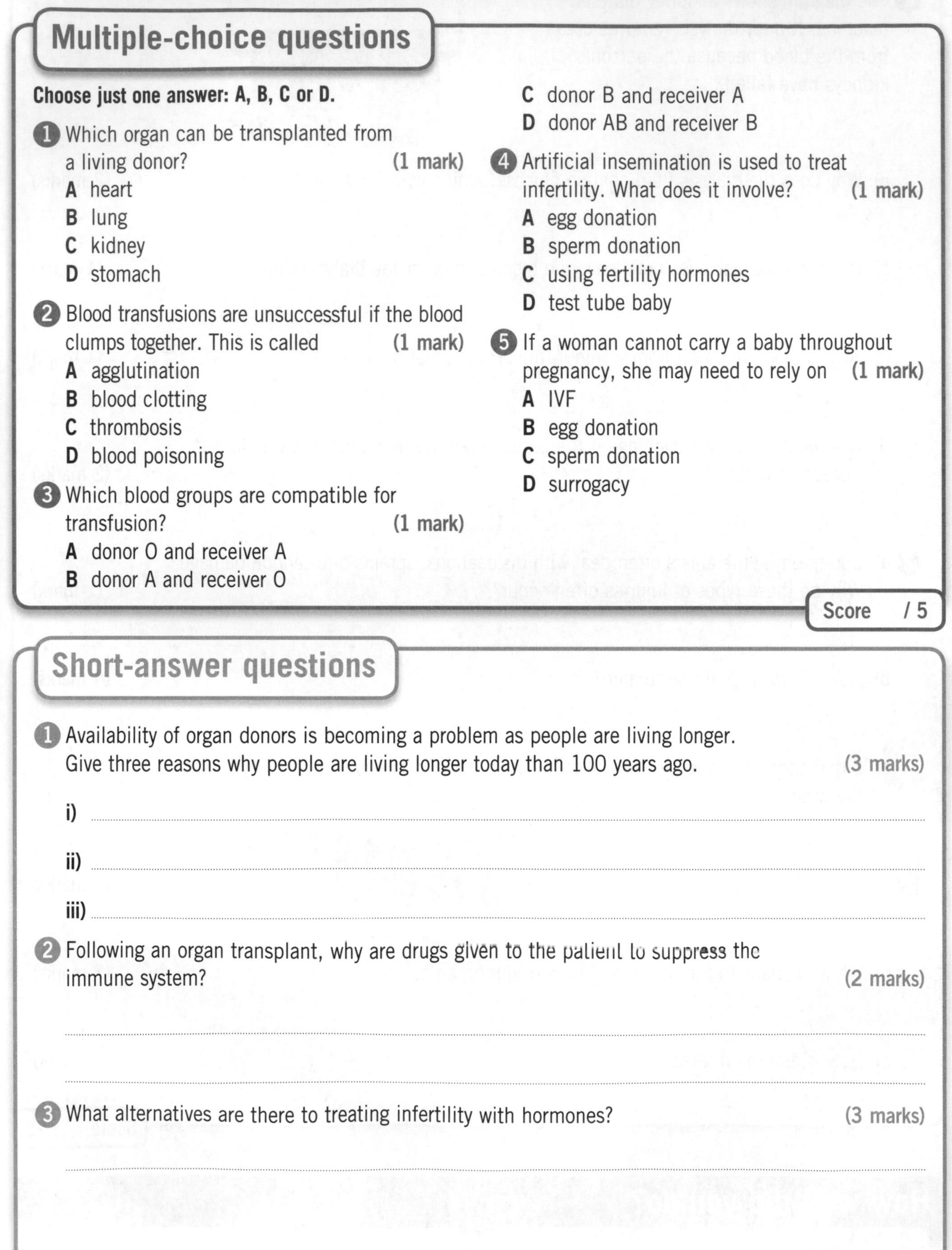

Multiple-choice questions

Choose just one answer: A, B, C or D.

1. Which organ can be transplanted from a living donor? **(1 mark)**
   - **A** heart
   - **B** lung
   - **C** kidney
   - **D** stomach

2. Blood transfusions are unsuccessful if the blood clumps together. This is called **(1 mark)**
   - **A** agglutination
   - **B** blood clotting
   - **C** thrombosis
   - **D** blood poisoning

3. Which blood groups are compatible for transfusion? **(1 mark)**
   - **A** donor O and receiver A
   - **B** donor A and receiver O
   - **C** donor B and receiver A
   - **D** donor AB and receiver B

4. Artificial insemination is used to treat infertility. What does it involve? **(1 mark)**
   - **A** egg donation
   - **B** sperm donation
   - **C** using fertility hormones
   - **D** test tube baby

5. If a woman cannot carry a baby throughout pregnancy, she may need to rely on **(1 mark)**
   - **A** IVF
   - **B** egg donation
   - **C** sperm donation
   - **D** surrogacy

Score /5

Short-answer questions

1. Availability of organ donors is becoming a problem as people are living longer. Give three reasons why people are living longer today than 100 years ago. **(3 marks)**

   i) ....

   ii) ....

   iii) ....

2. Following an organ transplant, why are drugs given to the patient to suppress the immune system? **(2 marks)**

   ....

   ....

3. What alternatives are there to treating infertility with hormones? **(3 marks)**

   ....

Score /8

GCSE-style questions

Answer all parts of all questions. Continue on a separate sheet of paper if necessary.

1. At present, people opt into a system to donate organs and carry a donor card. Some medical groups feel that it should be assumed that organs are available for transplant unless the person has opted out. Evaluate the reasons for and against a 'presumed consent' system. **(4 marks)**

2. **a)** In blood group system ABO, why is blood group O known as a universal donor? **(2 marks)**

   b) Explain why blood group A would not result in a successful transfusion if donated to a person with blood group B. **(3 marks)**

   c) What is agglutination? **(1 mark)**

3. Joint problems are often treated by resurfacing the ends of the bones. Which material would be better for this purpose, metal or plastic? Explain your answer. **(1 mark)**

Score / 11

How well did you do?

0–5	Try again	6–11	Getting there	12–17	Good work	18–24	Excellent!

For more information on this topic, see pages 102–103 of your Success Revision Guide.

Types of Behaviour

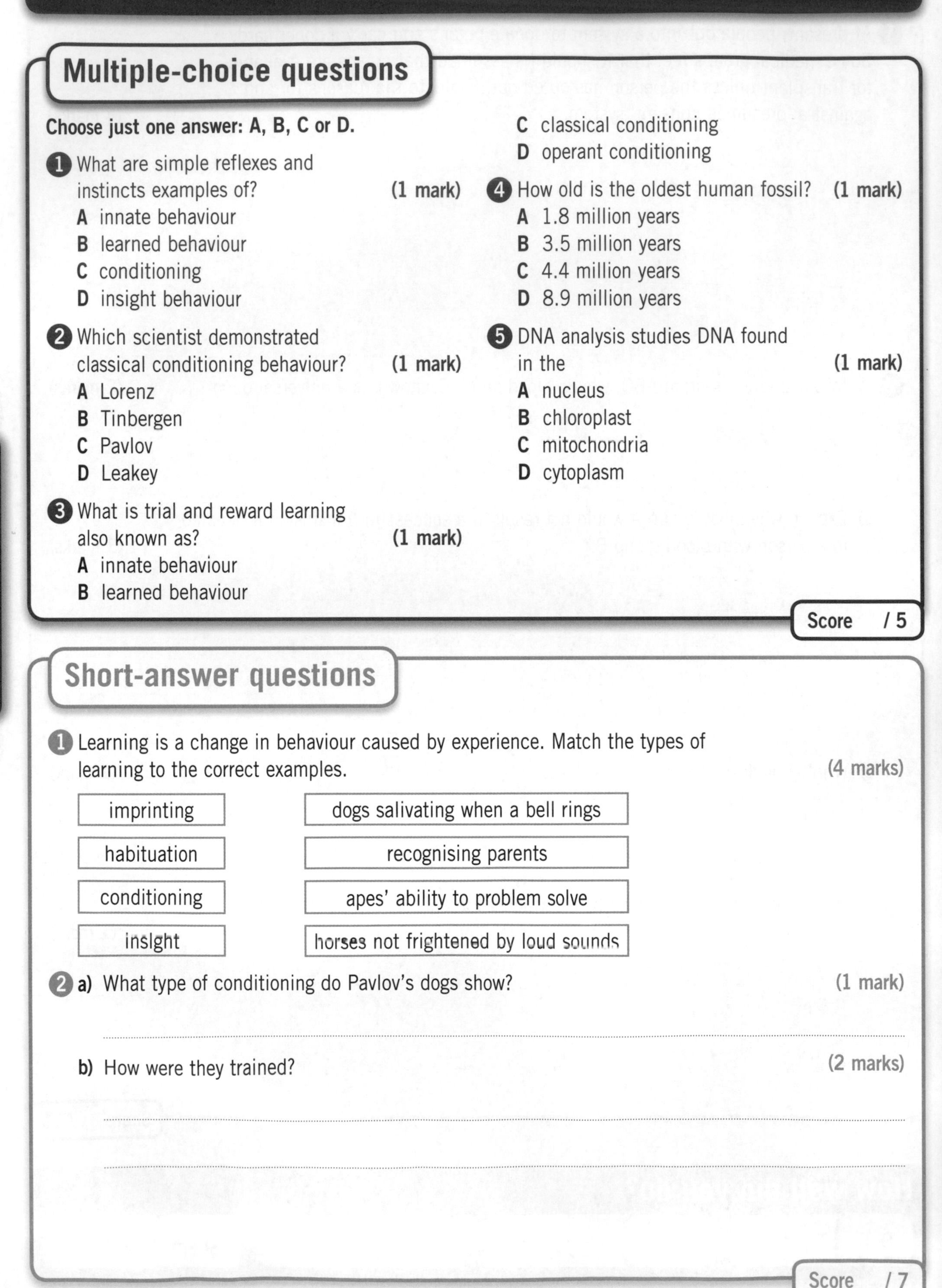

Multiple-choice questions

Choose just one answer: A, B, C or D.

1. What are simple reflexes and instincts examples of? **(1 mark)**
   A innate behaviour
   B learned behaviour
   C conditioning
   D insight behaviour

2. Which scientist demonstrated classical conditioning behaviour? **(1 mark)**
   A Lorenz
   B Tinbergen
   C Pavlov
   D Leakey

3. What is trial and reward learning also known as? **(1 mark)**
   A innate behaviour
   B learned behaviour
   C classical conditioning
   D operant conditioning

4. How old is the oldest human fossil? **(1 mark)**
   A 1.8 million years
   B 3.5 million years
   C 4.4 million years
   D 8.9 million years

5. DNA analysis studies DNA found in the **(1 mark)**
   A nucleus
   B chloroplast
   C mitochondria
   D cytoplasm

Score /5

Short-answer questions

1. Learning is a change in behaviour caused by experience. Match the types of learning to the correct examples. (4 marks)

imprinting	dogs salivating when a bell rings
habituation	recognising parents
conditioning	apes' ability to problem solve
insight	horses not frightened by loud sounds

2. a) What type of conditioning do Pavlov's dogs show? (1 mark)

   ....

   b) How were they trained? (2 marks)

   ....

   ....

Score /7

Animal Behaviour

GCSE-style questions

Answer all parts of all questions. Continue on a separate sheet of paper if necessary.

1. How does mitochondrial DNA differ from DNA found in the nucleus? **(1 mark)**

2. a) Which part of the brain is involved in forming memories? **(1 mark)**

   b) What else is this part of the brain involved in? **(2 marks)**

3. a) Give an example of how operant conditioning could be used to train circus horses. **(2 marks)**

   b) Why do some people disagree with this type of conditioning? **(1 mark)**

4. a) What is innate behaviour? **(1 mark)**

   b) Why are they essential to the survival of a species? **(1 mark)**

5. As humans have evolved, their brains have become more developed. What evidence has been used to support this statement? **(3 marks)**

Score / 12

How well did you do?

0–6	Try again	7–12	Getting there	13–18	Good work	19–24	Excellent!

For more information on this topic, see pages 106–107 of your Success Revision Guide.

Communication and Mating

Multiple-choice questions

Choose just one answer: A, B, C or D.

1. Which organism cares for its young? **(1 mark)**
   - **A** fish
   - **B** frog
   - **C** pigeon
   - **D** snake
2. Which bird has one mate for life? **(1 mark)**
   - **A** pigeon
   - **B** sparrow
   - **C** eagle
   - **D** albatross
3. Which of the following is not used by animals to communicate with each other? **(1 mark)**
   - **A** sounds
   - **B** electrical signals
   - **C** visual signals
   - **D** chemical signals
4. Humans are great apes. Which of the following do NOT belong to this group? **(1 mark)**
   - **A** gorillas
   - **B** orangutans
   - **C** chimpanzees
   - **D** monkeys
5. Which animals did Jane Goodall study? **(1 mark)**
   - **A** gorillas
   - **B** orangutans
   - **C** chimpanzees
   - **D** monkeys

Score /5

Short-answer questions

1. List three methods of communication used by organisms. **(3 marks)**

   i) ....

   ii) ....

   iii) ....

2. **True or false?** **(4 marks)**

	True	False
a) Body language is a universal way of communicating between all organisms.	☐	☐
b) Birds and mammals care for their young.	☐	☐
c) The instinct for mammals to suck on the breast is an example of innate behaviour.	☐	☐
d) An example of a great ape is a monkey.	☐	☐

3. What does anthropomorphism mean? **(2 marks)**

   ....

Score /9

GCSE-style questions

Answer all parts of all questions. Continue on a separate sheet of paper if necessary.

1

a) What type of behaviour is shown by male birds displaying colourful feathers? **(1 mark)**

....

b) Why is it important for a female to select a mate carefully? **(1 mark)**

....

2 a) How do baby birds ensure that their parents feed them? **(1 mark)**

....

b) What is the advantage of caring for the young? **(1 mark)**

....

3 a) Chimpanzees are great apes. Give two other examples. **(2 marks)**

....

b) What similarities have been discovered between chimpanzees and humans? **(2 marks)**

....

....

4 Why are facial expressions described as species-specific? **(2 marks)**

....

....

....

Score / 10

How well did you do?

0–6	Try again	7–12	Getting there	13–18	Good work	19–24	Excellent!

For more information on this topic, see pages 108–109 of your Success Revision Guide.

The Variety of Microbes

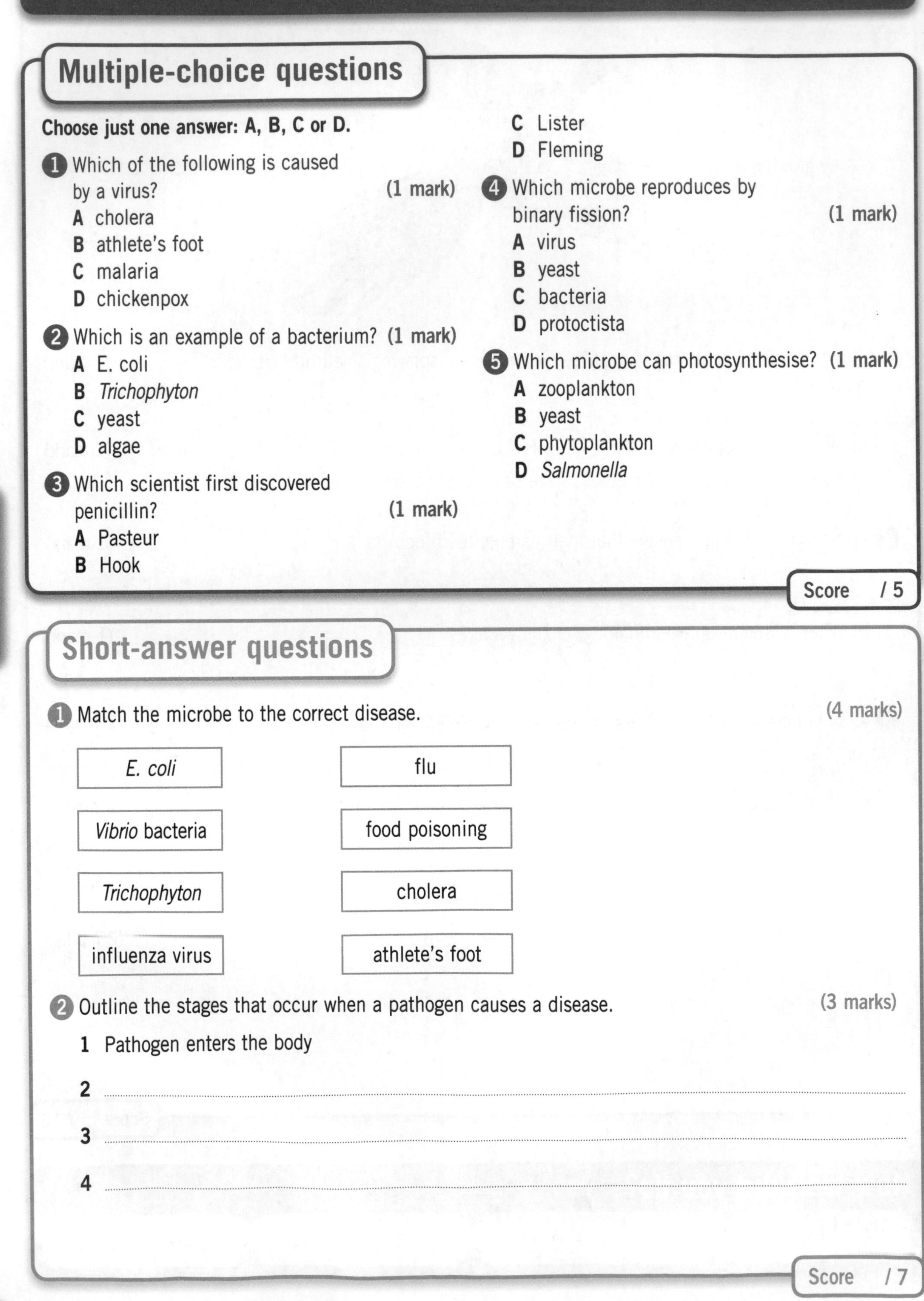

Multiple-choice questions

Choose just one answer: A, B, C or D.

1. Which of the following is caused by a virus? **(1 mark)**
   A cholera
   B athlete's foot
   C malaria
   D chickenpox

2. Which is an example of a bacterium? **(1 mark)**
   A E. coli
   B *Trichophyton*
   C yeast
   D algae

3. Which scientist first discovered penicillin? **(1 mark)**
   A Pasteur
   B Hook
   C Lister
   D Fleming

4. Which microbe reproduces by binary fission? **(1 mark)**
   A virus
   B yeast
   C bacteria
   D protoctista

5. Which microbe can photosynthesise? **(1 mark)**
   A zooplankton
   B yeast
   C phytoplankton
   D *Salmonella*

Score /5

Short-answer questions

1. Match the microbe to the correct disease. **(4 marks)**

Microbe	Disease
E. coli	flu
Vibrio bacteria	food poisoning
Trichophyton	cholera
influenza virus	athlete's foot

2. Outline the stages that occur when a pathogen causes a disease. **(3 marks)**

   1 Pathogen enters the body

   2 ....

   3 ....

   4 ....

Score /7

GCSE-style questions

Answer all parts of all questions. Continue on a separate sheet of paper if necessary.

1 Complete the sentences about microbes. **(8 marks)**

a) Yeast belongs to the .... kingdom.

b) Yeast is a .... celled organism, which reproduces .... by growing a bud on the side.

c) Bacteria cells have many different shapes, e.g. spherical, ... and spiral. Some bacteria can move using a ....

d) Viruses are simple structures consisting of a .... and ....

e) Protoctista includes single-celled ....

2 Modern medicine has enabled many infectious diseases to be treated and cured. Medicine and vaccinations are also used to prevent the spread of certain diseases. Many scientists' ideas and methods have been used. Discuss the work of Pasteur, Fleming and Lister, describing how they helped modern medicine. **(6 marks)**

....

....

....

....

3 a) Microscopic organisms found in the sea, called plankton, belong to which kingdom? **(1 mark)**

....

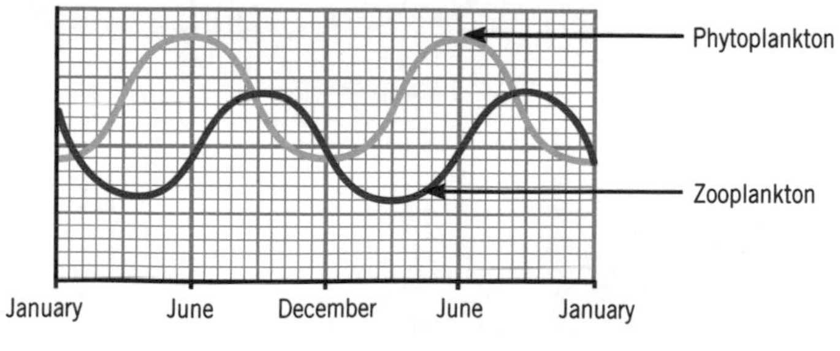

b) How do phytoplankton feed? .... **(1 mark)**

c) What factors affect the growth of phytoplankton? **(3 marks)**

....

Score / 19

How well did you do?

0–7	Try again	8–15	Getting there	16–23	Good work	24–31	Excellent!

For more information on this topic, see pages 112–113 of your Success Revision Guide.

Putting Microbes to Use

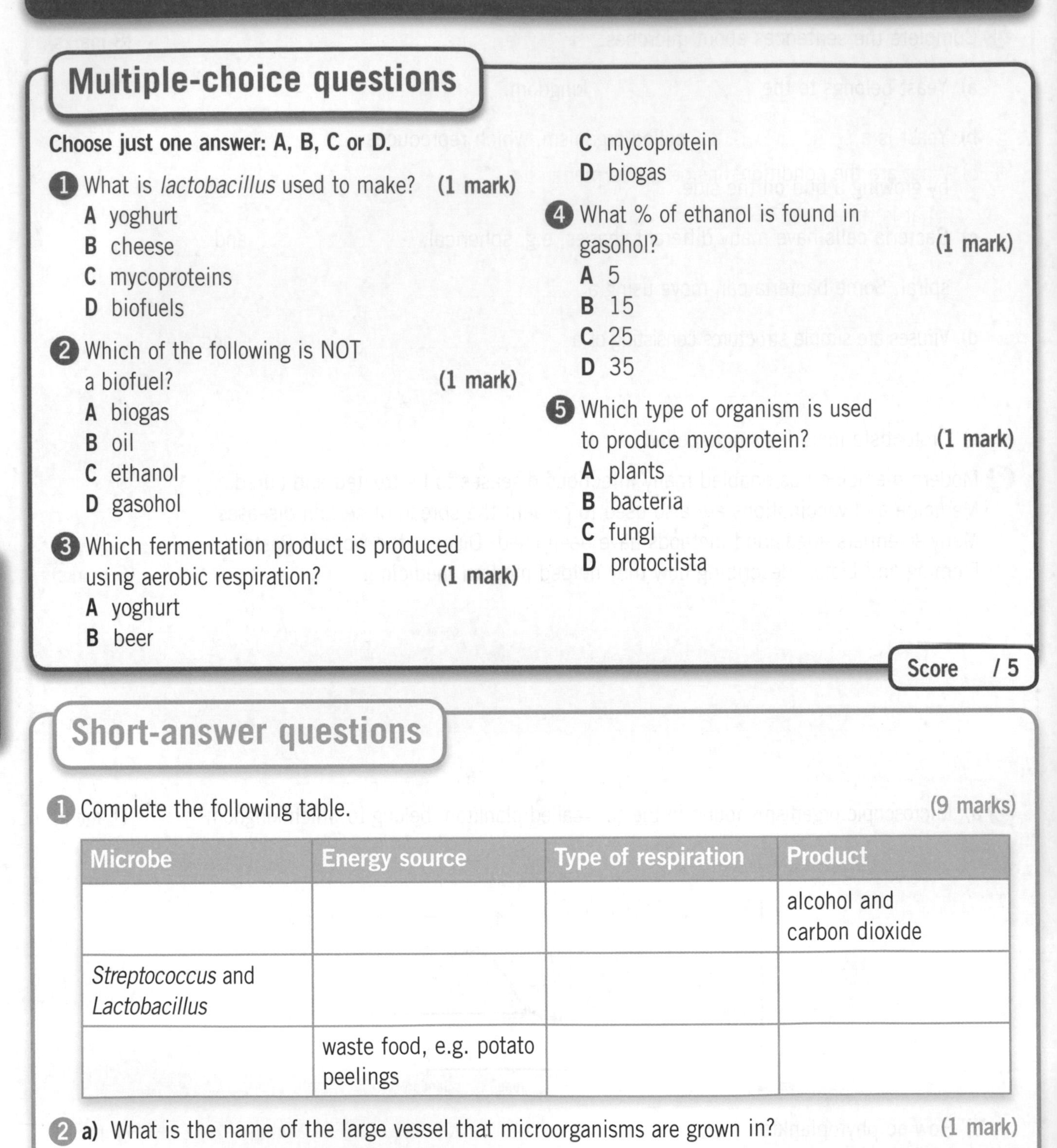

Multiple-choice questions

Choose just one answer: A, B, C or D.

1. What is *lactobacillus* used to make? **(1 mark)**
   - **A** yoghurt
   - **B** cheese
   - **C** mycoproteins
   - **D** biofuels

2. Which of the following is NOT a biofuel? **(1 mark)**
   - **A** biogas
   - **B** oil
   - **C** ethanol
   - **D** gasohol

3. Which fermentation product is produced using aerobic respiration? **(1 mark)**
   - **A** yoghurt
   - **B** beer
   - **C** mycoprotein
   - **D** biogas

4. What % of ethanol is found in gasohol? **(1 mark)**
   - **A** 5
   - **B** 15
   - **C** 25
   - **D** 35

5. Which type of organism is used to produce mycoprotein? **(1 mark)**
   - **A** plants
   - **B** bacteria
   - **C** fungi
   - **D** protoctista

Score /5

Short-answer questions

1. Complete the following table. (9 marks)

Microbe	Energy source	Type of respiration	Product
			alcohol and carbon dioxide
Streptococcus and *Lactobacillus*			
	waste food, e.g. potato peelings		

2. a) What is the name of the large vessel that microorganisms are grown in? (1 mark)

   ....

   b) What conditions are controlled in the vessel? (2 marks)

   ....

   ....

Score /12

GCSE-style questions

Answer all parts of all questions. Continue on a separate sheet of paper if necessary.

1 a) Why is all equipment sterilised before the bacteria are added to the milk when making yoghurt? **(1 mark)**

....

b) What are the conditions inside the fermenter? .... **(2 marks)**

c) What is the product of fermentation and what is its effect? **(2 marks)**

....

2 a) Write the word equation for the fermentation reaction involving yeast. **(2 marks)**

....

b) How are spirits produced? .... **(1 mark)**

c) Match the sugar source to the spirit. **(3 marks)**

sugar source	spirit
sugar cane	vodka
potatoes	whiskey
malted barley	rum

3 What are the advantages of mycoprotein as a high protein source compared to meat? **(3 marks)**

....

....

4 a) Define biofuels. .... **(1 mark)**

....

b) Biogas is an example of a biofuel. What is the name of the fermenter it

is produced in? .... **(1 mark)**

c) What does the biogas consist of? .... **(3 marks)**

d) What can biogas be used for? .... **(2 marks)**

....

Score /21

How well did you do?

0–9	Try again	10–19	Getting there	20–28	Good work	29–38	Excellent!

For more information on this topic, see pages 114–115 of your Success Revision Guide.

Microbes and Genetic Engineering

Multiple-choice questions

Choose just one answer: A, B, C or D.

1. Which enzymes cut DNA? **(1 mark)**
   A digestive
   B ligase
   C restriction
   D proteases

2. Which organism is now used to produce insulin? **(1 mark)**
   A sheep
   B cows
   C fungi
   D bacteria

3. Which is an example of a vector involved in genetic engineering? **(1 mark)**
   A plasmid
   B bacteria
   C mosquito
   D enzyme

4. Rice has been genetically modified to produce a vitamin. Which vitamin does it produce? **(1 mark)**
   A A
   B B complex
   C C
   D K

5. Which microorganism is used to produce chymosin? **(1 mark)**
   A virus
   B bacteria
   C fungi
   D protoctista

Score / 5

Short-answer questions

1. If an organism has been genetically modified it may be referred to as being 'transgenic'. What does this mean? (1 mark)

2. **True or false?** (5 marks)

	True	False
a) Ligase enzymes cut DNA.	☐	☐
b) *Agrobacterium tumefaciens* is a vector for transferring genetic material to fungi.	☐	☐
c) Cheese produced from chymosin obtained from GM microbes is suitable for vegetarians.	☐	☐
d) Genetic fingerprinting identifies the genes coding for certain proteins.	☐	☐
e) GM bacteria can produce insulin as they now contain an insulin molecule.	☐	☐

Score / 6

GCSE-style questions

Answer all parts of all questions. Continue on a separate sheet of paper if necessary.

1 Microbes, plants, animals and humans can all be genetically modified. Give an example of a possible use for each organism. **(4 marks)**

Microbe ....

Plant ....

Animal ....

Human ....

2 The following sentences describe how genetic engineering occurs. Put them in the correct order. **(6 marks)**

A DNA ligase joins the gene into the plasmid.

B The bacteria are grown in a fermenter.

C Restriction enzymes cut the DNA and plasmid.

D The gene is identified.

E The bacteria are tested to ensure they have taken up the plasmid containing the desired gene.

F The plasmid is removed from a bacterium cell.

3 **a)** Which bacteria is used as a vector to produce genetically modified plants? **(1 mark)**

....

b) Why is this bacteria referred to as a vector? **(1 mark)**

....

4 **a)** How is genetic fingerprinting used? **(2 marks)**

....

....

b) Outline the processes involved in genetic fingerprinting. **(3 marks)**

....

....

....

Score / 17

How well did you do?

0–7 Try again | 8–14 Getting there | 15–21 Good work | 22–28 Excellent!

For more information on this topic, see pages 116–117 of your Success Revision Guide.

Notes